T0332408

COVID-19 Public Health Measures

Intelligent Signal Processing and Data Analysis

Nilanjan Dey

*Department of Information Technology, Techno India College of
Technology, Kolkata, India*

Proposals for the series should be sent directly to one of the series
editors above or submitted to

Chapman & Hall/CRC
Taylor and Francis Group
3 Park Square, Milton Park
Abingdon, OX14 4RN, UK

For more information about this series, please visit: https://www.
routledge.com/Intelligent-Signal-Processing-and-Data-Analysis/
book-series/INSPDA

COVID-19 Public Health Measures

An Augmented Reality Perspective

Nuzhat F. Shaikh
Ajinkya Kunjir
Juveriya Shaikh
Parikshit Narendra Mahalle

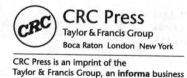

CRC Press
Taylor & Francis Group
Boca Raton London New York

CRC Press is an imprint of the
Taylor & Francis Group, an **informa** business

First edition published 2021
by CRC Press
6000 Broken Sound Parkway NW, Suite 300, Boca Raton, FL 33487-2742

and by CRC Press
2 Park Square, Milton Park, Abingdon, Oxon, OX14 4RN

© 2021 Nuzhat F. Shaikh, Ajinkya Kunjir, Juveriya Shaikh and Parikshit Narendra Mahalle

First edition published by CRC Press 2021

CRC Press is an imprint of Taylor & Francis Group, LLC

ISBN: 9781032003634 (hbk)
ISBN: 9781032003757 (pbk)
ISBN: 9781003173878 (ebk)

Typeset in Times
by codeMantra

DEDICATED TO ALL OF OUR
RESPECTIVE PARENTS

CONTENTS

PREFACE

We've all come across augmented reality at some point or the other in our lives. Be it while playing a game of Pokemon Go, amusing ourselves by clicking selfies trying out those wild Snapchat filters, decorating our homes through the IKEA app, or even while trying out different varieties of makeup with the L'Oreal app. That's augmented reality, to be present in the real world, yet to be able to interact with something that you can see and manipulate which isn't really there.

Work in AR has been increasingly receiving attention in the past decade. Despite the public fascination with augmented reality (AR), very few within the broader audience actually understand how these systems function. This book provides the basic aspects of Augmented Reality (AR) and the main concepts of this technology. The authors provide an interesting perspective to examine COVID-19 public health measures through AR. This book aims to examine how to use AR to help people change their behaviors (e.g., social distancing), which is an important dimension in curbing the current pandemic. However, in addition to examining "how" AR technologies can be applied in times of pandemic, a more important question is why it is important to use AR to help solve the problem. Looking at the evolving situation of coronavirus in various countries in this world, most number of COVID-19 outbreaks occurs in crowded locations such as schools, hospitals, and retail stores where there are more than 100 people present in an area all the time. The "SODAR" tool developed by Google for *social distancing* abides all the AR guidelines and displays a 6-feet distance circle around the person holding the mobile device. This tool can be used by people belonging to all kinds of categories, impaired and nonimpaired.

This book is organized and presented in such a way that an individual who has little or no knowledge will be able to develop AR apps after reading this book. The progression of the contents is such that new chapters build on the contents of the previous one. The first chapter begins with introducing the readers to the world of epidemic, endemic, and pandemics. Gradually, it acquaints the reader with the current situation that the large population across the globe is facing due to the COVID-19 pandemic. It also mentions the precautionary measures that can be taken to curb the spread of the pandemic and how AR can work as a supportive technology.

Chapter 2, "Augmented Reality Tools and Technology," sets the stage for further chapters by providing the definition of AR and a brief history of the development of this technology. It introduces the reader to the core components of AR—processor, displays, and sensors. It also deals with the computer vision methods that can be employed in AR. We have discussed the various software tools that are available to assist the reader in developing AR applications. AR is usually seen as a preferred technology for gaming and entertainment; in the last section, we bring light to on how AR can provide support and assistance to specially abled individuals.

The retail industry use case presented in Chapter 3 of this book follows an empirical approach by focusing on the "disabled" users due to their vulnerability towards the virus. It has been proven that the virus is the most active and attracted to either older or very young hosts, i.e., age groups above 50 or below 20 due to immunity systems. The main purpose of this AR application is to showcase an extra layer of information for all the products such as nutrition facts, price, availability on shelves, and price comparison with other stores. Having these features available in the application will reduce the chances of people needing help from other people (associates, random people, cashiers, etc.) and therefore avoiding COVID-19 reachability. The chapter also discusses the software tools such as Unity, Vuforia, and C# for AR application.

Chapter 4 discusses the marker-based AR technique in depth along with its advantages and disadvantages. It goes ahead to give the step-wise description for marker detection comprising Image Pre-processing, Discarding Outliers in the Marker, Deciphering of Markers and Calculation of the Marker pose. It further explains in detail the AR application design for social distancing model along with the proposed system architecture.

Chapter 5 mentions the mitigation remedies and recommendations. It throws light on some of the existing applications for social distancing such as Google SODAR, Aarogya Setu, 1point5, DROR, and Canada COVID-19. Chapter 6 ends with providing AR solution to the public health COVID-19 mitigation remedies and some useful recommendations over existing applications.

Chapter 6 summarizes this book by discussing the issues and challenges and future outlook. It also gives a list of the latest developments in the area of augmented reality, in the form of bibliography, which have played a very important role in getting together the contents of this book.

This book includes a glossary of some typical terms and terminologies which serve as a dictionary for the reader to reference throughout their reading of this book.

Appendix A gives the reader a few online sources for getting free AR tools.

Appendix B provides coding guidelines for deploying an AR app over an Android phone.

ACKNOWLEDGMENTS

It is a pleasure to thank all those who made this book possible. First and foremost, we would like to express our deepest gratitude to all the reviewers, who took out time to help us improve the quality of contents by raising very apt concerns during the proposal phase.

We owe an enormous debt of gratitude to those who gave us detailed and constructive comments on one or more chapters. They gave us a lot of their time to discuss nuances of the text and pushed us to clarify concepts, explore particular facets of insight work, and explain the rationales for specific recommendations. In doing so, they have surely helped us improve the quality of the books by making the contents richer and richer and thus improving the quality. We would specially like to thank our respective parents who have taken all the efforts to raise us and get us to a position to enable us to author a book. Their efforts cannot be ignored.

We cannot forget to mention Dr. Gagandeep Singh, Publisher (Engineering), CRC Press, for helping us at each stage from the proposal through to the submission of the final manuscript. He took out time to answer all our queries with patience and in the shortest time.

Our special gratitude to Dr. Nilanjan Dey, the series editor, for providing valuable guidance and suggestions.

AUTHORS

Dr. Nuzhat F. Shaikh obtained her bachelor's in Computer Engineering from Cummins College of Engineering for Women, and ME in Computer Engineering from Pune Institute of Computer Technology, SP Pune University, Pune, India. She completed PhD in Computer Science and Engineering from SGGSIET, Swami Ramanand Teerth Marathwada University, Nanded, India. She has more than 23 years of teaching and research experience. At present, she is working as Professor and Head, Department of Computer Engineering at Modern Education Society's College of Engineering (Wadia College), Pune.

Her research areas include artificial intelligence, soft computing, bio-inspired optimization methods, augmented and virtual reality. She is a reviewer of the Science and Information (SAI) Organization's *International Journal of Advanced Computer Science and Applications* (IJACSA), Elsevier journal of *Applied Soft Computing* (ASOC), and many more reputed journals. She has published more than 40 technical papers at various conferences and in reputed journals. She has also remained a technical program committee member and session chair for many international conferences.

She is a recognized PhD guide of SSPU, Pune, and is guiding 4 PhD students in the area of AI, IoT, and machine learning. She has authored a book on multimedia techniques.

She is a life member of the CSI and the ISTE. She has been conferred the "Educator Excellence Award 2018" by Odser Charitable Trust for her contribution to the teaching fraternity.

Mr. Ajinkya Kunjir obtained his BE degree in Computer Engineering from Modern Education Society's College of Engineering, India. He completed his master's in computer science from Lakehead University, Ontario, Canada in the year 2021. He was a junior QC Engineer at Ubisoft, Pune, India. He has completed his full-time coop program at Ministry of Education, Toronto, Canada. Currently, he is a reviewer for IGI Global—*International Journal of Big Data and Analytics in Healthcare* (IJBDAH), and also serves as an ad-hoc reviewer of other IGI journals such as IJHIS and IJTICS. The areas of interest for research are data mining, machine learning, compiler design, and Big Data technology. Other than core research interests, other computer science fields studied and passed are game programming, artificial intelligence, web health informatics, deep learning, and ethical issues in computer science.

Ms. Juveriya Shaikh obtained her BE degree in Computer Engineering from Modern Education Society's College of Engineering, SP Pune University, India. She is working at Persistent Systems, Pune, as a software engineer. Her expertise includes machine learning, artificial intelligence, data science, Internet of Things, augmented reality, and mixed reality. She has developed multiple applications for education and interior designers using Microsoft Hololens. She has good practical knowledge of Unity and Vuforia.

Dr. Parikshit Narendra Mahalle obtained his BE degree in Computer Science and Engineering from Sant Gadge Baba Amravati University, Amravati, India, and ME degree in Computer Engineering from Savitribai Phule Pune University, Pune, India. He completed his PhD in Computer Science and Engineering specialization in wireless communication from Aalborg University, Aalborg, Denmark. He was Post Doc Researcher at CMI, Aalborg University, Copenhagen, Denmark. Currently, he is working as Professor and Head in the Department of Computer Engineering at STES's Smt. Kashibai Navale College of Engineering, Pune, India. He has more than 20 years of teaching and research experience. He is serving as a subject expert in Computer Engineering, Research and Recognition Committee at several universities like SPPU (Pune), and SGBU (Amravati).

He is a senior member of IEEE, ACM member, life member of CSI, and life member of ISTE. Also, he is a member of *IEEE Transactions on Information Forensics and Security, IEEE Internet of Things Journal.* He is a reviewer for IGI Global—*International Journal of Rough Sets and Data Analysis* (IJRSDA), Associate Editor for IGI Global—*International Journal of Synthetic Emotions* (IJSE), and Inderscience *International Journal of Grid and Utility Computing* (IJGUC). He is a Member-Editorial Review Board for IGI Global—International Journal of Ambient Computing and Intelligence (IJACI). He is also working as an Associate Editor for IGI Global—*International Journal of Synthetic Emotions* (IJSE). He has also remained a technical program committee member for international conferences and symposiums like IEEE ICC, IEEE INDICON, IEEE GCWSN, IEEE ICCUBEA, etc.

He is a reviewer for the Springer journal of *Wireless Personal Communications*, reviewer for Elsevier journal of *Applied Computing and Informatics*, member of the Editorial Review Board of IGI Global—*International Journal of Ambient Computing and Intelligence* (IJACI), and member of the Editorial Review Board for *Journal of Global Research in Computer Science.*

He has published more than 150 research publications having 1103 citations and H index of 13. He has five edited books to his credit published by Springer and CRC Press. He has seven patents to his credit. He has also delivered invited talk on "Identity Management in IoT" to Symantec Research Lab, Mountain View, California. He has delivered more than 100 lectures at national and international level on IoT, Big Data, and Digitization. He has authored 13 books on subjects like Context-aware Pervasive Systems and Application (Springer Nature Press), Design and Analysis of Algorithms (Cambridge University), Data Analytics for COVID-19 Outbreak by CRC Press, Identity Management for the Internet of Things (River Publications), Data Structure and Algorithms (Cengage Publications), Programming using Python (Tech-Neo Publications MSBTE)

He had worked as Chairman of Board of Studies (Information Technology), SPPU, Pune. He is working as Member—Board of Studies (Computer Engineering), SPPU, Pune. He has been a member of the Board of Studies at several institutions like VIT (Pune), Government College of Engineering (Karad), Sandeep University (Nashik), Vishwakarma University (Pune), Dr. D. Y. Patil International University (Pune), etc. He has also remained a technical program committee member for many international conferences.

He is a recognized PhD guide of SSPU, Pune, and guiding seven PhD students in the area of IoT and Machine Learning. Recently, two students have successfully defended their PhD, and he has research finding of more than 1 million to his credit. He is also the recipient of "Best Faculty Award" by Sinhgad Institutes and Cognizant Technologies Solutions. His recent research interests include Algorithms, Internet of Things, Identity Management, and Security.

PANDEMIC AND GLOBAL OUTBREAKS

1.1 INTRODUCTION

Pandemics and global outbreaks have always been a threat to not only mankind but also the wildlife risking their population to get endangered. Pandemics have stopped human life numerous times and have struck the world hard a number of times and claimed many human lives.

As indicated by the Centers for Disease Control and Prevention (CDC)—a public general well-being establishment in the United States—"Epidemiology is the study of the distribution and determinants of health-related states or events in specified populations and the application of this study to the control of health problems" [1].

Since a lot of news has emerged recently regarding dengue, Zika virus (2016), H1N1 (2009), SARS, Hanta virus, and most recently COVID-19 (2019), it has become more significant to understand the difference between the epidemiological terms such as epidemic, endemic, pandemic, and outbreak. Spanish flu, common cold, SARS, MERS, and H1N1 are few of the viruses that have been able to invade the human population and claim a number of lives. We have yet again been in the hard times of facing the difficult situation of fighting the battle against another global pandemic, COVID-19. As per the weekly epidemiological update published by the World Health Organization (WHO), more than 53.1 million people have been affected by the infectious coronavirus disease. The virus has killed more than 1.3 million humans since the start of the pandemic until November 14, 2020. Not only does the battle stop at preventing ourself from getting infected but also to be able to invent and discover a vaccine or treatment against the virus in the soonest of time.

At the time of writing, we are almost 11 months into 2020 and the world is struggling with this existential global crisis. The WHO called the COVID-19 viral disease a pandemic. The governments across the world advised their people to follow safety guidelines in order to help prevent the spread of the virus. Being a contagious and a touch-invoked infection, the virus can quickly capture a host just by sitting next to or by talking to an infected person from a short distance. The epicenter of this global outbreak was learned to be at Wuhan, China, and a strategy was formed to contain the spread on an epidemic level. The range of virus being widespread and impacts being dangerous, the virus crossed country borders and was declared as a pandemic. The world is experiencing difficulties and is also following the precaution guidelines and staying safe.

1.2 PANDEMIC AND EPIDEMIC: AN OVERVIEW

An epidemic is named a circumstance wherein an infection is spreading actively. Epidemic can be termed or defined as a phenomenon where thousands of people in a region are infected by a deadly infection or virus. As against this, the term pandemic identifies with geographic spread and is utilized as a phrasing to depict an infection which has influenced an entire nation or the total globe. A global epidemic is termed as a pandemic. It is a plague that has spread its wings more than a few nations or landmasses of the globe influencing countless individuals. At the point when the quantity of cases surpasses more than what might be normal, it is named an epidemic, in epidemiologic terms, similar to the COVID-19. If there is an outbreak that affects most of the world, it is referred to as a pandemic. It generally describes an unexpected increase in the number of infected people. An outbreak can occur in a particular community, geographical region, or several countries at the same time.

The term endemic is used when there is a disease that exists ceaselessly inside a geographic area. Endemic contamination is about infections, microscopic organisms, and microorganisms that exist inside a particular area topographically. While informal use of the term epidemic may not require such distinction, it's critical to be able to differentiate between the two terms (and some identical ones like *endemic* and *outbreak*) when dealing with news related to public health.

Epidemiological terms that would be used in this book are

- **Outbreak:** It is used if the number of infections exceeds more than what would be expected

- **Endemic:** A contamination/infection within a geographic area that is existing continually.
- **Epidemic:** A situation when a disease is spreading actively.
- **Pandemic:** A global endemic

1.2.1 Endemic

The term endemic refers to a sickness, infection, or well-being worry that is continually present in a given area or populace inside a geographic district. The quantity of affected people is low and doesn't essentially increase or reduce after some time; For example, chicken pox, which influences kids at sporadic unsurprising rate in the UK and jungle fever or malaria, which is continually present in numerous places of Africa. In the event that we think about the case of dengue fever, there are numerous parts on the planet where dengue fever has been proclaimed as an endemic by WHO, which implies that there are mosquitoes existing in these regions that are transmitting the dengue fever and communicating it from individual to individual. In any case, imported cases and imported flare-ups of dengue are additionally found in various regions where this infection is definitely not an endemic. Recently in 2015–2016, there was a flare-up in the Big Island of Hawaii where somebody, probably came in affected with dengue fever, got bit by mosquitoes, and subsequently there was a local chain of transmissions where those mosquitoes then bit others, they got dengue fever, and so on. For this circumstance, dengue fever isn't endemic in the Big Island; however, there was an outbreak because of an imported infection with resulting transmission.

1.2.2 Epidemic

Thucydides, an Athenian officer in B.C. times described Plague of Athens to be an earliest occurrence of an epidemic disease. Until the 17th century, endemic and epidemic were two terms, which were considered to depict contrasting conditions of diseases based on the level of population being affected. Endemic depicted a condition at low rates of occurrence and the epidemic was considered as a widespread condition. Any problem that grows out of control or extends beyond an intended population is termed as an epidemic. Some previous examples of such epidemic problems are the 1918 Spanish flu, measles outbreak, and 2014's whooping cough.

Atlanta's Center for Disease Control [2] describes an epidemic as

> the occurrence of more cases of the disease, injury, or other health condition than expected in a given area or among a specific group of persons during a particular period. Usually, the cases are presumed to have a common cause or to be related to one another in some way.

An epidemic holds a reputation of spreading rapidly in a contained area, e.g., the seasonal flu, which sometimes is referred to as an endemic, behaves more like an epidemic because cases spike at a certain time of the year. Some classical epidemics in history include the yellow fever which killed 5,000 people in Philadelphia in 1793 and Cholera in the United States which killed over 1,50,000 Americans in three waves. It is still prevalent and fatal in many parts of the world.

An epidemic disease may or may not be infectious. WHO has used the term for both the West Nile fever, which is infectious, and to obesity, which is noninfectious.

The two significant conditions which lead to the flare-up of epidemics may include movement of populations of certain animals, such as rats or mosquitoes, which may act as illness transporters and the supply of infected food such as contaminated drinking water.

There are a few epidemics which may be governed by certain seasons. For instance, people suffer from whooping cough during springs, while measles results in two epidemics, one in March and the other one in winter. Flu, common cold, and different diseases of the upper respiratory tract, for example, sore throat, occur predominatingly in the winter.

There is another variety of epidemic, both with respect to the quantity of individuals being influenced by the epidemic and the quantity of individuals who die in successive epidemics, common source outbreak, and propagated outbreak.

- **Common source outbreak:** The affected individuals had an exposure to a common transmitter in a common source outbreak epidemic. On the contrary, if the exposure is singular and all of the affected people develop the disease as a result of exposure to a single source and incubation course, it very well may be named a point source outbreak, but if the exposure is nonstop or variable, it can be termed an intermittent or continuous outbreak.
- **Propagated outbreak:** The disease spreads person-to-person in a propagated outbreak, and the individuals who are affected

may become independent reservoirs which may lead to further exposures. At most of the times, epidemics will display characteristics of both common source and propagated outbreaks, which are very often referred to as a mixed outbreak.

1.2.3 Pandemic

An epidemic becomes a pandemic when it affects a large percent of the population, crosses national and international boundaries, and, most of the times, affects people on a global scale. Essentially, a pandemic is not limited to a specific geography or population. It usually affects a significant number of people across countries and continents. In short, pandemic is an upgrade of epidemic at an international or massive population level. A disease or a medical condition cannot be termed as a pandemic just because it has spread over a large area or it has killed numerous people, but it must be infectious too. As an example, cancer is responsible for claiming thousands of lives, but it is not regarded as a pandemic because the disease is neither contagious nor infectious.

The most prolific pandemics from history include the HIV/AIDS that has claimed the lives of more than 36 million people since 1981, and over 35 million cases are still prevalent in the world: the Spanish flu in 1918, which infected a population of 500 million, with an estimated mortality of 10%–20% and the bubonic plague, which spread across Europe, Asia, and Africa, killing an estimated 75–200 million people, to name a few.

Pandemic can be distinguished from the rest of the epidemiological terms due to the following reasons:

- In addition to a large population being affected, it spreads worldwide on a geographical area.
- Human beings have little to no immunity against the virus, and hence it spreads at a very fast rate.
- It is caused by a deadly virus that hasn't infected anyone since a long period of time.
- It results in more deaths.
- Most of the time, it creates economic loss and huge social disruption.

1.3 COVID-19 OUTBREAK

On December 31, 2019, different instances of pneumonia or influenza-like cases brought about by obscure reasons were distinguished in Wuhan City and reported to the WHO China Country Office in Hubei

Province of China. Patients were isolated, and the health authorities started work on tracing the source of the flu.

Within a span of 4 days, on January 3, 2020, a total of 44 patients with pneumonia of unrealized origin had been reported to the WHO by the national authorities of China. The market in Wuhan which was suspected to be where the infection began was shut on January 1, 2020, for ecological sterilization and cleansing. The organic microbe that caused the contamination was not yet distinguished or affirmed.

The symptomatic signs and side effects were identified as fever, with a few patients experiencing breathing difficulties and * chest radiographs demonstrating invasive lesions of both the lungs.

As identified by the authorities, recognized patients included wholesalers, dealers, or merchants from the Huanan seafood market. Scientists were of the view that, taking everything into account, there was a high possibility that the contamination started from bats but first went through a go-between creature similar to another coronavirus—the 2002 SARS outbreak—that moved from horseshoe bats to cat-like civets prior to influencing individuals. In any case, there was vulnerability around a few parts of the beginning story that researchers were trying hard to unwind, including which species passed it to a human, since identifying how a pandemic starts is a fundamental perspective for ending the following one.

This was how the first few cases of the then epidemic were reported. The preceding paragraphs explain the sequence of events that took place after the initial cases until the discovery of the cause of the cases and then finally WHO deciding the name that would be used to address the virus. Presented below is a timeline of the events that have taken place.

- **January 6–7, 2020:** Introductory assessments concerning the outbreak ruled out bird flu, occasional influenza, SARS, and MERS. The Chinese authorities had uncovered that the quantity of suspected cases touched almost 60 with seven being in a critical condition. Human-to-human transmission had not been affirmed nor been denied by the health authorities. Chinese authorities had perceived the contamination, called coronavirus, which is a group of viruses including the common cold, SARS, and MERS. Incidentally, this new disease was named 2019-nCoV.
- **January 11–12, 2020:** China shared the genetic sequence of the novel coronavirus. This was a very significant step, as it

provided the others with the much-needed information to develop precise diagnostic kits. A case of the novel coronavirus was confirmed in Thailand, by health officials, on January 13, 2020. It was expected that cases of the novel coronavirus were bound to emerge outside of China in a short duration of time, and WHO had called for active checking and readiness in other countries.

- **January 22–23, 2020:** The WHO Director-General brought together the members of the Emergency Committee to talk about the outbreak of the novel coronavirus in China, with few cases additionally being reported in the Republic of Korea, Thailand, Japan, and Singapore.

  Considering the prohibitive and paired nature of the new pandemic, the vast majority of the committee individuals thought of it as still too soon to announce a Public Health Emergency of International Concern (PHEIC). It was incomprehensible as of now that the infection would cause such immense destruction over the globe, within a short time.

- **January 30, 2020:** WHO Director-General Tedros Adhanom Ghebreyesus announced the 2019-nCoV outbreak a public health crisis of global concern, taking note of the expected spread of the infection to nations with frail health systems. The decision was taken since numerous nations outside China reported instances of contamination. The two countries, Philippines and India, affirmed their first 2019-nCoV cases. China pronounced that the absolute confirmed cases had arrived at 9,692, with 213 casualties. WHO suggested "2019-nCoV acute respiratory disease" as the interim name for the illness.

- **February 11, 2020:** Rules ordered that the name of the disease could not refer to a geological area, a creature, an individual, or a gathering of individuals. It was additionally expected to identify with the sickness and be pronounceable. That is how the WHO named the infection as Covid-19 and tweeted, "We currently have a name for the #2019nCoV illness: COVID-19."

- **March 11, 2020:** WHO had proclaimed the novel COVID-19 outbreak a worldwide pandemic. At a press conference, WHO Director-General, Dr. Tedros Adhanom Ghebreyesus, noticed that in the course of a few weeks, the count of cases outside China increased 13-fold, and the count of countries with cases tripled.

- **April 4, 2020:** WHO had affirmed that around 1 million instances of COVID-19 had been recognized over the globe, a more-than-ten-times increment in under a month. WHO had released new technical guidance that recommended universal access to public hand hygiene stations and making their use mandatory on entering and leaving any public or private commercial building and any public transport facility.

After this, until May 28, 2020, there were a total of 5.69 million confirmed cases out of which 2.35 million had recovered and 356 thousand deaths had been recorded. By this time, the countries affected included China, South Korea, Thailand, Japan, the United States, Hong Kong, Malaysia, Australia, Cambodia, Germany and Sri Lanka, Russia, Sweden, Spain, India, and Philippines.

- **August 30, 2020:** India recorded a single-day spike of 78,761 infections taking India's COVID-19 total tally to more than 35 lakhs, with the count of deaths crossing 63,000, as updated by the Union Health Ministry of India. This was the highest single-day spike for any country across the globe. India had left behind its own record of 77,266 infections in a single day on August 27, 2020 (Table 1.1).

- **September 11, 2020:** India reports 96,760 new Covid cases in a single day—the biggest single-day spike by any nation since the beginning of the pandemic. A half year since WHO proclaimed the worldwide pandemic, India's COVID-19 count crossed the 45-lakh mark. India has added 96,760 new coronavirus infections—the most noteworthy single-day spike any nation has ever included in 24 hours. India's demises in a single day have likewise crossed that of the US, which is the most noticeably terribly hit by the pandemic. There were 1,213 deaths announced in 24 hours, which is likewise the most elevated casualty revealed in a single day.

TABLE 1.1
Top 5 Highest Single-Day Spikes, Country-wise

Country	No. of Cases	Date
India	78,761	August 29, 2020
India	77,266	August 27, 2020
USA	77,255	March 21, 2020
Brazil	69,074	July 29, 2020
Chile	36,179	June 17, 2020

Source: Johns Hopkins University data.

- **October 24, 2020:** The US announced its most elevated number of COVID-19 contaminations in a single day since the pandemic's beginning. The US reported in excess of 80,000 new coronavirus infections on Friday, October 23, 2020—the most elevated day-by-day case number since the pandemic started.
- **November 12, 2020:** Researchers globally are working nonstop to discover an antibody against SARS-CoV-2, the virus causing the COVID-19 pandemic. The Herculean efforts imply that an optimized immunization could arrive at market anyplace from the finish of 2020 to the center of 2021. Specialists have raised extensive worry about the antibody's security and adequacy given it has not yet entered Phase 3 clinical preliminaries. Until the present, only two coronavirus vaccines have been approved. Sputnik V—once known as Gam-COVID-Vac and created by the Gamaleya Research Institute in Moscow—was approved by the Ministry of Health of the Russian Federation on August 11, 2020. Serum Institute of India, the world's biggest vaccine maker, said that it has made 40 million portions of AstraZeneca's potential COVID-19 vaccine (AZD1222), co-created by Oxford University, and would before long start making Novavax's opponent shot, as they both look for administrative approval.
- **November 14, 2020:** Globally, there have been 53,164,803 confirmed instances of COVID-19, including 1,300,576 deaths, as reported by WHO (Figures 1.1–1.3; Table 1.2).

1.4 ORIGIN OF THE VIRUS

The coronavirus disease 19 (COVID-19) is an exceptionally communicable and pathogenic viral contamination caused by severe acute respiratory syndrome coronavirus 2 (SARS-CoV-2), which arose in Wuhan, China and spread far and wide. Genomic investigation uncovered that SARS-CoV-2 is phylogenetically related to severe acute respiratory syndrome-like (SARS-like) bat infections, and accordingly, bats could be the conceivable essential store. The intermediate source of origin and transfer to people isn't known; however, the speedy human-to-human transmission has been greatly confirmed. According to a study published by Scripps Research Institute in Science Daily on March 17, 2020, Andersen and his associates reasoned that the most probable inceptions for SARS-CoV-2 followed one of two potential situations (Figure 1.4).

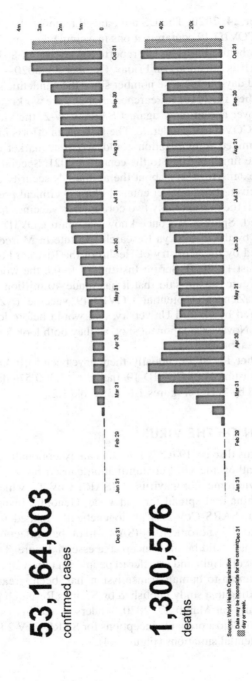

Figure 1.1 Variation occurrences in the amount of confirmed positives and deaths. (https:// covid19.who.int/)

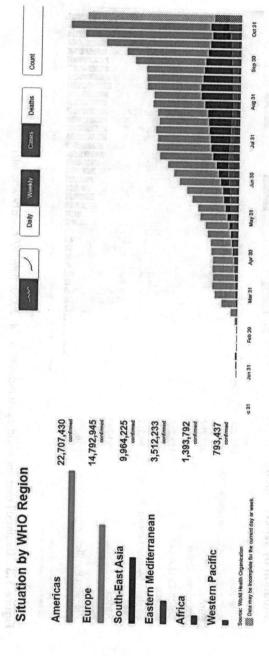

Figure 1.2 Confirmed cases all over the world as on November 14, 2020. (*https:// covid19.who.int/*)

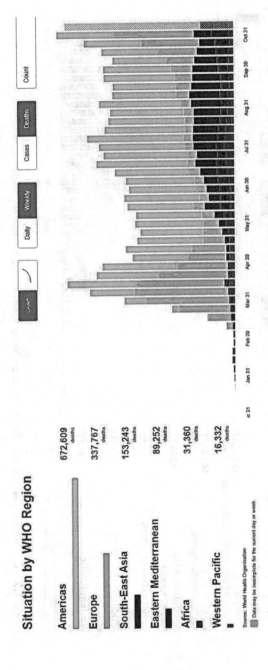

Figure 1.3   Deaths all over the world as on November 14, 2020. (https:// covid19.who.int/)

TABLE 1.2
COVID-19 Statistics Worldwide

Confirmed cases	54,373,119
Active cases	15,149,634
Recovered	37,904,531
Deceased	1,318,954

November 15, 2020, 08:54 GMT, https://www.worldometers.info/coronavirus.

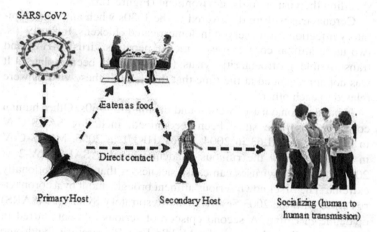

Figure 1.4 Transmission of coronavirus to humans.

In one situation, the infection developed to its present pathogenic state through characteristic choice in a nonhuman host and afterward leaped to people. This is the means by which past coronavirus outbreaks have arisen, with people getting the infection after direct exposure to civets (SARS) and camels (MERS). The scientists proposed bats as the most probable store for SARS-CoV-2 as it is fundamentally the same as a bat coronavirus.

In the other conceivable situation, a nonpathogenic adaptation of the infection bounced from an animal to humans and afterward developed to its present pathogenic state inside the human population. For example, some coronaviruses from pangolins, armadillo-like mammals found in Asia and Africa, have a RBD structure, fundamentally the same as that of SARS-CoV-2. A coronavirus from a pangolin might have been transmitted to a human, either straightforwardly or through a delegate host, for example, civets or ferrets.

1.5 MEDICAL OVERVIEW OF COVID-19

Coronaviruses are a large family of viruses that belong to the Coronaviridae family. Corona represents the crown-like spikes on the outer surface of the virus; thus, it was named coronavirus. These viruses are minute in size, ranging from 65 to 125 nm in diameter, and contain a single-stranded RNA. The spikes are 20 nm long. Infection begins when the viral spike protein of the virus attaches to its complementary host cell receptor. Infected carriers are capable of shedding the virus into the environment (Figure 1.5).

Coronaviruses were discovered in the 1930s when an acute respiratory infection was observed in domesticated chickens. In the 1940s, two more animal coronaviruses, mouse hepatitis virus (MHV) and transmissible gastroenteritis virus (TGEV), had been isolated. It was not apprehended at the time that the three of these viruses were related to each other.

Human coronaviruses were found during the 1960s. Other human coronaviruses have since been recognized, including SARS-CoV in 2003, HCoV NL63 in 2004, HCoV HKU1 in 2005, MERS-CoV in 2012, and now the ruinous pandemic causing SARS-CoV-2 of 2019. The Coronaviruses can cause sicknesses that are exceptionally extreme. The first known serious ailment brought about by a coronavirus arose with the 2003 Severe Acute Respiratory Syndrome (SARS) epidemic in China. A second episode of serious ailment started in 2012 in Saudi Arabia with the Middle East Respiratory Syndrome (MERS). On December 31, 2019, Chinese authorities alarmed WHO a flare-up of a novel strain of coronavirus causing serious disease, which was subsequently named SARS-CoV-2.

1.5.1 Background of SARSv2, Spanish Flu, H1N1, and COVID-19

Coronaviruses differ essentially in the danger factor. A few like the MERS-CoV can kill over 30% of those affected, and some like the normal virus causing common cold are comparatively harmless. Coronaviruses can cause colds with major symptoms, for example, fever, sore throat, etc.

Three human Coronaviruses produce conceivably extreme symptoms:

1. Middle East respiratory syndrome-related coronavirus (MERS-CoV), β-CoV

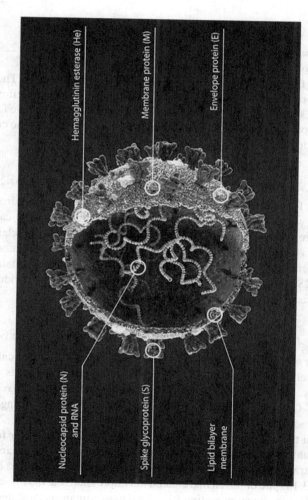

Figure 1.5 Bisectional view of coronavirus. (https://img.theweek.in)

2. Severe acute respiratory syndrome coronavirus (SARS-CoV), β-CoV
3. Severe acute respiratory syndrome coronavirus 2 (SARS-CoV-2), β-CoV

1.5.2 Common Cold

The human coronaviruses HCoV-OC43, HCoV-HKU1, HCoV-229E, and HCoV-NL63 constantly stream in the human populace and produce the generally mild symptoms of the common cold. These coronaviruses cause about 15% of regular colds, while the leftover 40%–50% of colds are brought about by rhinoviruses. The above four mild coronaviruses have an occasional occurrence usually in the cold weather months.

1.5.3 Severe Acute Respiratory Syndrome (SARS)

In 2003, following the outbreak of severe acute respiratory syndrome (SARS) which had started a year prior in Asia, and auxiliary cases elsewhere on the planet, WHO gave an official statement proclaiming that a novel coronavirus recognized by a few research laboratories was the causative agent for SARS. The infection was formally named the SARS coronavirus (SARS-CoV), and in excess, 8,000 individuals were infected, around 10% of whom died.

1.5.4 Middle East Respiratory Syndrome (MERS)

Another sort of coronavirus was recognized, which was at first called the Novel Coronavirus 2012 in September 2012 and afterwards was formally named Middle East respiratory syndrome coronavirus (MERS-CoV). A worldwide alarm was given by WHO regarding the severity of MERS-CoV. The WHO update on September 28, 2012 said that the infection didn't appear to communicate effectively from individual to individual. Notwithstanding, instances of human-to-human transmission in France and Tunisia were affirmed around May 12, 2013. In spite of this, it had created the impression that the infection faced inconvenience spreading from human to human, as most people who were infected didn't communicate the infection. The virus was given another name, Human Coronavirus—Erasmus Medical Center (HCoV-EMC) after the Dutch Erasmus Medical Center sequenced the virus. The last name for the infection was then finally announced to be Middle East respiratory syndrome coronavirus (MERS-CoV).

As of December 2019, 2,468 instances of MERS-CoV contamination had been affirmed out of which 851 were lethal and a death percentage of around 34.5%.

1.5.5 Spanish Flu

The Spanish influenza pandemic of 1918, the deadliest ever, affected around 500 million individuals around the world—which is around 33% of the planet's populace—and had taken the lives of 20–50 million of those infected by the virus. The 1918 influenza was first seen in Europe, the United States, and parts of Asia before it quickly started spreading far and wide. No powerful medications or vaccinations to treat this deadly influenza strain were accessible around then. Citizens were ordered to wear masks, schools, theaters, and organizations were closed, and bodies had accumulated in mortuaries before the infection finished its fatal worldwide walk.

Flu, or Spanish influenza or just influenza as it is most commonly referred to, is an infection that attacks the respiratory framework. Being profoundly infectious, when an infected individual coughs, sneezes, or talks, respiratory droplets are formed and communicated into the air and would then be breathed in by anybody in the region. Moreover, an individual who touches something with the infection on it and afterwards contacts their mouth, eyes, or nose would also get infected.

1.5.6 Swine Flu or H1N1 Flu

Swine flu was a disease that was brought about by the H1N1 virus. A strain of swine flu called H1N1 affected numerous individuals around the globe in 2009. The virus is infectious and can communicate from human to human. Symptoms of swine flu in individuals are like that of standard human influenza and incorporate fever, cough, sore throat, body pain, migraine, chills, and tiredness. Antiviral prescriptions can prevent or treat swine flu. There is a vaccine accessible to protect against swine flu.

1.5.7 Coronavirus Disease 2019 (COVID-19)

A pneumonia outbreak was accounted for in Wuhan, China in December 2019. On December 31, 2019, the outbreak was followed by a novel strain of coronavirus, which was given the interim name of 2019-nCoV by WHO, which was afterwards renamed SARS-CoV-2 by the International Committee on Taxonomy of Viruses.

Respiratory diseases can be transmitted through various-sized droplets. As indicated by studies and current proof, the COVID-19 infection is essentially transmitted between individuals through contact and respiratory droplets. Droplet transmission happens when an individual is in close contact, i.e., inside 1 m (approx.) with somebody who has respiratory indications like coughing or sneezing and is consequently in danger of having his/her mucosae (mouth and nose) or conjunctiva (eyes) presented to the infective respiratory drops. Transmission of the COVID-19 infection can likewise happen by direct contact with infected individuals and indirect contact with surfaces in the immediate environment or with objects utilized by or on the infected individual (e.g., stethoscope or thermometer). The COVID-19 virus is sufficiently huge to be communicated through the air as it settles down because of its weight.

1.6 PREVENTION OF PANDEMICS

Arrangements and measures that are powerful as a reaction to a pandemic consist of various stages. The main stage is having a disease surveillance framework, for example, the capacity to quickly and urgently send emergency health workers, particularly locally based emergency health workers, and a legitimate and passable approach to guarantee that the security and wellbeing of these health officials is dealt with; e.g., there is a public lab in Tanzania that runs testing for, in excess of 200 health sites and screens, the spread of irresistible infections. The following stage is the real and pragmatic reaction to a crisis. Michael Gerson, a U.S.-based journalist had expressed in 2015 that only NATO and the U.S. military have the worldwide ability to react to such a crisis, where infectious illnesses are concerned. But, notwithstanding the broadest preparatory measures, a quick spreading pandemic may effortlessly overpower the current health care resources. Hence, early and proactive mitigation efforts, whose essential point should be the so-called "epidemic curve flattening," should be taken. Typically, such measures comprise nonpharmacological intercessions, for example, social/physical separating, decided contact tracing, "stay-at-home" or lockdown orders, as well as appropriate personal protective equipment, for example, gloves, masks, protective garments for health workers, and other physical barriers to stop the spread of the pandemic.

A significant errand in dealing with the outbreak of an infectious disease is putting forth an attempt to decrease the epidemic peak, known as "flattening the epidemic curve." This helps in decreasing the risk of health services being affected and provides adequate time for a vaccine to be developed. In an influenza pandemic, these activities may include personal preventive measures, for example, hand cleanliness, wearing face masks, and self-isolation; community measures which focus on social distancing, for example, shutting down schools/universities/commercial centers and not permitting get-togethers and natural measures; for example, sterilizing and disinfecting surfaces that are utilized frequently.

Containment and mitigation are the two fundamental methodologies that should be attempted in the control of a pandemic. Containment can be embraced in the initial phases of the outbreak. It tends to be accomplished by isolating the infected people to prevent the sickness from spreading to the remainder of the populace. Some other public health interventions on disease control and remedial counter measures, for example, vaccinations, may also be effective if available [3].

The mitigation stage should be executed, when it gets clear that it's not possible to forestall the spread of the infection. In this stage, moves are made to hinder the spread of the sickness and moderate its impacts on society and furthermore the medical care framework. Essentially, containment and mitigation measures should be undertaken simultaneously.

The significant objectives of mitigation incorporate deferring and diminishing peak burden on medical care, i.e., flattening the curve and taking measures to decrease the overall cases and health impact. This strategy gives sufficient time to the administrators to progressively increase the healthcare capacity such as bed count, personnel, and equipment.

Another procedure, suppression, requires more extreme long-term nonpharmaceutical interventions to turn around the pandemic by diminishing the basic duplication number to under 1. The suppression procedure incorporates ground-breaking lockdown measures, populace-wide social distancing, home isolation of active cases, and family isolation and quarantine. These measures are being taken by most of the COVID-19-affected countries including India during the pandemic where all of the cities had been placed under strict lockdown for a few months. The lockdown, at present, is being gradually lifted taken into consideration all the precautionary measures. A few countries that did not impose strict lockdown or rather their population did not accept the lockdown have seen a second wave of the pandemic.

To stop the spread of COVID-19, there have been multiple measures suggested by WHO. The guidelines included practices such as social distancing, wearing masks and hand gloves, using sanitizers frequently, and maintaining a 6-feet distance from people. Considering the overall aspect of the situation, this book focuses on the use of augmented reality (AR) applications for preventing COVID-19 outbreak along with techniques, tools, and platforms that will be of assistance for users and help achieve social distancing and sanitization.

The motivation for this book arose because of the lack of the following required precautions due to essential human needs and also because of inevitability of the disabled users to stand the technology and safety guidelines. It is important for normal human beings to follow the precautions and also to understand and assist the disabled users with the right technology to follow the safety guidelines and stay updated with them for avoiding further spread of the virus. In response to address these issues and challenges, AR, artificial intelligence, and Internet of Things (IoT) will play an important role in achieving public safety measures upfront.

This book will focus on the theoretical and practical knowledge of AR and on how to use this technology to prevent the spread of COVID-19. This book also contains multiple use cases along with a set of recommendations. Another focus of this book will be on building applications using opensource software with an interactive interface to aid the impaired users. The initial pages will emphasize on the basic knowledge of AR, the technology, devices, and rest of the relevant theories. Two aspects which need to be considered when using AR technology are usability (knowledge on how to use the tool) and flexibility (procedures to follow when and where). This book will also provide extra resources and coding aid for practitioners planning to build AR applications for public health safety measures.

2

AUGMENTED REALITY TOOLS
AND TECHNOLOGY

2.1 AUGMENTED REALITY OVERVIEW

Augmented reality (AR) is rapidly growing in the recent years and capturing the attention of the masses. Most likely, many of us have been experiencing AR without realizing it. Two of the most liked social media apps, Snapchat and Instagram, have many filters that are used in the user's posts and stories. AR technology is employed by all these filters, which we have been using unknowingly every day.

This chapter makes an attempt to explain the meaning and idea of the term AR in the context of this book. At the same time, it will also introduce some terms that will aid in understanding the chapters that follow.

2.1.1 What Is AR?

AR creates automatic, direct, and actionable links between the electronic information and the physical world. It also provides an immediate and simple user interface to a physical world that is electronically enhanced. AR is capable of overlaying the real-world views with computer-generated information, enhancing human cognition and perception in remarkable new ways. Even though AR has made major developments, people are yet not aware of how AR augments reality and what superimposition means.

The most widely accepted definition of AR was proposed by Azuma in his survey paper published in 1997. According to the author [4], an AR system must consist of the following three characteristics:

- It must combine real and virtual world
- It must be interactive in real-time
- It must be registered in 3D

This definition doesn't limit itself to a particular output device, for example, a head-mounted display (HMD). Neither does it limit AR to only visual media. Haptic, gustatory, or olfactory and audio AR media have also been covered in its scope. The definition requires spatial registration and real-time control, i.e., accurate real-time alignment of corresponding real and virtual data.

Opinions on what is termed as real-time performance would differ from person to person and would also depend on the application, whereas interactivity implies that the human-computer interface operates in a tightly coupled feedback loop. This implies the AR scene is continuously navigated by the user, and the user also controls the AR experience, and in turn, the system fetches the input by tracking the user's position. The system then registers virtual content and the position in the real world and then presents a visualization that is registered to objects in the real world. At least three major components are required by a complete AR system: a registration component, a visualization component, and a tracking component. A fourth component, i.e., a spatial model (i.e. a database) is used to store data about the virtual and real world. Real-world model serves as a reference to the tracking component, which determines the user's location in the real world. The content required for augmentation is present in the virtual-world model. Both parts of the spatial model must be registered in the same coordinate system.

2.1.2 History of AR

In 1990, the term "augmented reality" was authored by Boeing researcher Tim Caudell, although Ivan Sutherland's development of the first head-mounted display system had marked the invention of AR technology way back in 1968.

The innovation has made some amazing progress since 1990, with the rundown of use cases developing as time passes. From NASA simulations to immersive advertising encounters, AR makes errands simpler—and certainly more fun!

The accompanying course of events attempts to clarify the way that the AR innovation has taken to go to its present position.

1968: Ivan Sutherland, a professor at Harvard and a renowned computer scientist, invented the first head-mounted display called "The Sword of Damocles."

1974: A lab called "Videoplace" was developed at the University of Connecticut that was completely committed to artificial reality by Myron Kruger, a computer researcher and artist.

1978: The idea of creating a virtual on-field marker to help TV viewers identify first down distances, in a football game, was conceived and patented by David W. Crain in 1978 [5].

1990: The term "AR" was coined by a Boeing researcher, Tom Caudell.

1992: One of the first fully functional AR systems "Virtual Fixtures" was created by Louis Rosenburg, a researcher in the USAF Armstrong's Research Lab.

1994: Julie Martin, a writer and producer, brought AR to the entertainment industry for the first time with the theater production titled *Dancing in Cyberspace*.

1998: The first live NFL game with the virtual 1st & Ten graphic system – aka the yellow yard marker was broadcasted by Sportsvision. 1st & Ten is a computer program that augmented the televised coverage of American football game by virtually inserting graphical elements on the field of play as if they were physically present [6]. The technology displayed a yellow line overlayed on top of the feed so that viewers can quickly see where the team just advanced to get the first down.

1999: NASA created a hybrid synthetic vision system of their X-38 spacecraft. The system leveraged AR technology to assist in providing better navigation during their test flights.

2000: An opensource software library, ARToolKit, was developed by Hirokazu Kato. This package helped other developers build AR software programs. The SDK uses video tracking to overlay virtual graphics on top of the real world.

2003: Sportsvision enhanced the First & Ten graphic to include the feature on the new Skycam system—it provided viewers with an aerial shot of the field with graphics virtually overlaid on top of it.

2009: *Esquire Magazine* made an attempt to make its pages come alive by using AR in print media for the first time.

2013: The MARTA app (Mobile Augmented Reality Technical Assistance) was introduced by Volkswagen primarily. Technicians were given step-by-step repair instructions by this app, within the service manual.

2014: A major breakthrough was the Google Glass device unveiled by Google. It is a pair of AR glasses that provides users with an immersive experience.

2016: Microsoft started shipping its version of wearable AR technology called the HoloLens, which was much more advanced than the Google Glass, but the price was very high.

2016: Pokemon Go, the very popular location-based AR game, was launched by Niantic and Nintendo. Pokemon Go has put AR on the mainstream map.

2017: Introduction of Apple's ARKit and Google's ARCore software development kits (SDKs)

2017: IKEA, the Swedish furniture company, released its AR app, called IKEA Place that changed the retail industry forever. Ikea Place represents one of the AR's most prominent use cases. The app uses AR to bring its sofas, shelves, and tables to life in a living room.

2019: Google launched a beta version of its AR walking direction feature for Google Maps that will be available to all AR-compatible iOS and Android mobile devices.

2020: According to a report from Gartner, at least 100 million users are expected to utilize AR-enabled shopping technologies by 2020, which happens to be one of the hottest retail trends of this year.

It has been predicted that by the end of 2021, 25%–30% of the people would prefer working from home for many days in a week. Microsoft is moving forward with a beta of a video-calling system which would employ AR to create holograms of participants. Cisco Systems is also working on a project called Musion that would bring together its networking products with AR technologies.

2.1.3 How AR Works?

AR transforms the environment around us into an advanced digital interface by putting virtual objects into real world, in real time. AR browsers enhance camera display with the contextual information. AR adds digital content onto a live camera feed, making that digital content look as if it were a part of the real world. For example, if the smart phone is pointed toward a product in the retail store, AR can be employed to display its stock in the store or its estimated cost or its nutrients.

There are many other forms of AR, e.g., auditory or haptics, but we would focus on the mainstream applications of AR, i.e., visual.

In general, two primary things need to take place for every time step in any AR application. The two-step process of any AR application is:

1. The application needs to determine the current state of the real and virtual world.

2. The application then needs to display the virtual world in registration with the real world in a manner such that the participant(s) are able to sense the virtual world elements as a part of his or her real world and then return to step 1 to move on to the next time step.

Obviously, there are a number of sub-steps in each of the above steps, but at the core of what is happening, these are the two key steps to be focused on right now. On considering the above steps, you can readily see that many different methods could be used to achieve them, and many different technologies could be used to implement the methods.

2.2 AR HARDWARE

AR is one of the most promising technologies at present. After the innovation and revolution of smartphones, everybody mandatorily carries one with them all the time, and all these mobile devices contain of a processor, GPS, a display, camera, microphone etc., which is all the hardware required for AR as well.

2.2.1 Displays

The display is a device that provides the signals that our senses perceive. Displays provide signals to our eyes, our ears, our sense of touch, and our nose, and perhaps provide a sensation of taste. Additionally, some displays provide stimuli designed to cause other sensations, such as to our vestibular system. This book focuses primarily on displays for our eyes and ears and provides a much lighter treatment of other types of displays.

This section is organized as follows:

- Visual displays
- Audio displays
- Haptic displays
- Other sensory displays
- Stereo displays (stereoscopic and stereophonic)

2.2.1.1 Visual Displays

The primary role of visual displays is to create signals of light that are perceived as visual images by our eyes. We are all familiar with the typical desktop computer monitor which is a classic example of visual display. It shows digital information from a computer in a way that our eyes can see and our brains can perceive as visual images.

The primary classes of visual displays used for AR applications include the following:

- **Stationary visual displays:** Stationary visual displays are displays that, as their name implies, do not move during their typical use. They are placed in a position, and they remain there. Of course, they can be moved to a new location, but, like a typical television set, you don't typically carry it with you wherever you go. In other words, you must go to the display to use it. A typical desktop computer monitor/LCD/LED would fall into the category of a stationary display.

- **Visual displays that move with the participant's head:** Some AR systems use displays that are mobile and move with the participant's head and are sometimes referred to as head-mounted displays (HMDs). The most common instantiations of these are displays that are worn like helmets, glasses, or headphones (audio), with the trend being toward lightweight glasses. Some displays could be considered hybrids, in that they move with the participant's head, but overall are stationary displays in which the participant places his head at the display and moves the display with his or her head in a limited range of motions.

- **Visual displays that move with the participant's hand or other parts of their body:** Currently, the most prevalent type of display for AR applications moves with the participant's hand. This is due to the widespread adoption of smartphones (such as the iPhone and Android phones) and smart tablets (such as the iPad and similar devices). Many developers recognize that a huge fraction of the population carries sensors, processors, and display devices in their pocket or purse regularly. Hence, even though they are more limited in some regard (processing, memory, screen size, etc.), their portability and ease of use make them a prime target for AR applications.

2.2.1.2 Audio Displays

The second most common sensory display is the audio or sonic display. Analogous to visual displays, display signals that our eyes can sense, sonic displays produce audio signals that our ears can sense. Our eyes and our ears convey different types of information to our bodies to help us perceive the world around us. Similar to visual displays, audio displays can also be classified into the following categories:

- Audio displays that move with the participant's head
- Stationary audio displays
- Audio displays that move with the participant's hand or other parts of his or her body

Stationary Audio Displays: AR displays are not limited to only visual presentations. The second most common sensory stimulus displayed is audio. The audio version of a stationary display is a stationary loudspeaker system, commonly referred to simply as "speakers." In general, home stereo speakers are an example of a stationary audio display.

Audio Displays That Move with the Participant's Head: Audio displays that move with the participant's head fall into two primary forms:

- Headphones
- Earbuds

Both headphones and earbuds are devices, which fall in the category of audio-to-sound transducers. One important distinction is the way they are worn. Headphones are typically worn over the ear(s), whereas earbuds are worn inside the ear(s). This distinction can affect how people perceive the sounds they are hearing. Earbuds take more advantage of sound traveling through our skeletal system in the head, whereas headphones are more limited to their impact on our eardrums. Headphones and earbuds overcome many sound pollution problems and also offer more privacy to the participants wearing them. People who are quite close to the participant do not hear what the participant is hearing unless the volume of the sound is quite loud. Likewise, the person wearing the headphones or earbuds is unlikely to hear most sounds around them unless they are exceedingly loud.

Audio Displays That Move with the Participant's Hand or Other Parts of His or Her Body: Smartphones, smart tablets, and other handheld devices often have an audio display built in. They often also have a jack available to allow you to plug in a pair of headphones or earbuds. The key aspect of handheld audio displays is that they can be either private or not private depending on the participant's wishes. That is, with the volume turned up, it can be heard by anyone in the area within the limits of the available volume. Or the volume can be turned down and the display held against one ear. Finally, for complete privacy, the participant can plug in a pair of headphones or earbuds.

2.2.1.3 Other Sensory Displays

In addition to sight and sound, AR applications may display the other human senses as well. The most common of these are the haptic (touch) and olfaction (smell) senses.

2.2.1.4 Haptics

Haptics, in general, refers to our sense of touch. This can roughly be broken into two components:

- Skin sensations (taction)
- Forces (kinesthetics)

Skin sensations include things such as temperature, texture, and pain. Forces are sensed by how our body responds to them. For example, when you lift a rolling ball, your body must overcome the force of gravity. Based on how much various muscles must react to do this, your body gains an understanding of the weight of the ball. Your body's sense of proprioception enables your mind to understand your bodily configuration even when your eyes are closed. As you move your hand over the ball, the muscles in your arms and fingers report information about its shape, and skin sensations tell you that the ball is smooth and cool.

A device such as the PHANTOM Omni from Geomagic is used to "feel" the object. The device is an articulated arm with a pen-type device on the end. You move the pen around to probe into space. If the device was in registration with an AR visual display, you could see the virtual object on the table and feel it by touching it with the pen. You could gain a sense of the shape of the object in much the same way as you could gain information about the shape of a real object by probing it with a pen.

2.2.1.5 Smell (olfaction)

Smell can be used as an aid in AR applications. One thing that makes the use of smell problematic is that it is difficult to "clear" a smell rapidly when it should go away. Smell is generally administered globally as a mist from an atomizer under computer control or more specifically to a particular person via tubing. There are use cases where a particular smell is used to indicate danger. For example, certain gases have been characterized by fragrances, although are very harmful, but have no noticeable smell. A good example can be seen in mines where very sharp fragrances have been added to the air supply to indicate danger and the need to evacuate the mine [7].

In general, smells are preformulated as a specific smell that is administered rather than a more general model where a set of scent "primitives" are combined to create new smells dynamically under the control of an algorithm in a way that would be analogous to using color theory to use red, green, and blue primitives (or another model) to create any color for display.

It is nontrivial to dispense smells/perfumes very rapidly, especially in large areas. It is even more difficult to "clear" an area of a scent that has been dispensed there. Hence, it is difficult to provide an environment in which different scents can be placed and eliminated rapidly.

2.2.2 Processors

A processor is an important component of an AR system. Note that the term processor can refer to a single processor unit or multiple components working in conjunction to provide a processing system. Processors in AR systems fulfill several roles. Processor is the "brain" of the system (of course, as mentioned previously, the magic of AR happens in the brain of the participant). Therefore, the basic roles of the processor are to receive signals from the sensors, execute the instructions from the application program based on the sensor information, and create the signals that drive the display(s) of the system. Computing systems for AR can range in complexity from simple handheld devices such as tablets and smartphones to desktop computers, laptops, and workstations and all the way to powerful distributed systems. In all cases, a computer must have enough computational capability to perform the tasks it needs to in *real time*. By real time we mean that every time any action is made, such as a button press or a change in orientation of a handheld device or change in point of view, the system should respond with an updated display of the combination of the virtual and real world with no delay.

In general, a processing system in AR systems consists mainly of one or more general-purpose microprocessors such as the central processing unit (CPU) and may or may not consist of one or more special purposes, i.e., graphics processing units (GPUs). GPUs are hardware that are specially optimized for executing 3D graphics computations. Since 3D graphics are one of the most common outputs from any AR system, most of the AR applications are expected to execute intensive image processing algorithms to detect particular objects on which virtual information is to be overlaid. Object detection generally consists of a feature description module and a feature extraction module. GPUs are often used to improve the performance of these methods [8].

2.2.2.1 Processor System Architectures

Several configurations/architectures are used for AR applications. Some of the common architectures for the processing functions are as follows:

1. Application run on a handheld system such as a smartphone
2. Application run on desktop/laptop computer
3. Application run as a web application
4. Application run on a cloud with a thin client
5. Other combinations of local and remote systems

2.2.2.2 Processor Specifications

Several specifications indicate roughly how a processing system will work for a given application. Some of the more important specifications include the following:

1. Number of processors
2. Network latency
3. Network bandwidth
4. Available memory
5. Processor speed
6. Graphics accelerator(s)

2.2.3 Tracking and Sensors

Sensors receive information about the physical world and communicate it to the AR application for a variety of purposes, one of which is to provide information about the location and orientation of the participant (or a surrogate for the participant, such as a handheld device like a smartphone) to the AR application. For example, through the use of sensors, the application can determine where the participant is and his or her pose in the real world. Other roles of sensors include providing information such as temperature, pH, lightness/darkness, or any other types of information about the environment to the AR application.

Tracking: Although there are many ways to do tracking in AR applications, by far, the most commonly used technique, especially for indoor applications, is through computer vision, which is an example of one way to do optical tracking. The specific sensor used for optical tracking is a camera. The camera gathers light through a lens and provides a signal that represents an image of what the camera "sees." That image is then analyzed to determine the desired tracking information.

In much the same manner that cameras can be used as sensors in optical tracking systems, microphones can be used as sensors in acoustical tracking systems. The analogy follows in that the microphone(s) can be attached to the object being tracked or can be placed in the environment. With acoustical tracking systems, there needs to be a source of acoustical information to be sensed by the microphones. Generally, ultrasound is used. Ultrasound is a sound that is higher in frequency than the human hearing system can perceive. In general, ultrasonic tracking systems work by having the object being tracked emit sound and an array of microphones in the environment to capture that sound. Based on the timing and amplitude of sound sensed in each microphone, one can compute the location of the source of the sound. Different objects can emit different frequencies of sound to track multiple objects in the same space. Conversely, objects can be tracked by having microphones attached to the objects and sound sources in the environment. In this case, however, the objects being tracked would need to be connected to the computing system to report the signals their microphones are sensing.

Another alternative to optical tracking is electromagnetic tracking. Because electromagnetic tracking systems are often utilized in virtual reality systems, they are well understood by those in that community, and there is software that is readily available to use such systems. Electromagnetic tracking systems can track in six degrees of freedom. The most common way they are implemented is that there is a transmitter with three orthogonal antennas. There is a corresponding receiver attached to the entity one wishes to track which also has three orthogonal antennas. The sensor is the receiving unit. The transmitter emits a signal, sequentially through each of its antennas. The signal acquired by each of the antennas is then analyzed, and the level of signal reported by each antenna can be used to compute the location and orientation of the receiver.

Another method of tracking, mechanical tracking, operates by attaching linkages to the object one wishes to track. These linkages have sensors at each of the joints that report the angle between the linkages. Often, this is done by placing a variable resistor (potentiometer) at the joint and reading the voltage there. As the angle of the linkage changes, the amount of resistance in the potentiometer changes, and a corresponding change in the voltage (that one can measure) occurs. The voltage can then be used to determine the angle between linkages. This information, in combination with the angles between all other linkages in the system, can be used to compute the location and pose of the object.

Sensors for tracking: The so-called depth sensors can be used for tracking in AR applications. The term depth sensor is used in several different technologies that can be used to provide information about how far away an object is from the sensor. The underlying technologies can be optical, acoustical (ultrasound), radar, etc. The result from a depth sensor is a measurement of how far an object is from the depth sensor. This information can be used in conjunction with other tracking technologies to provide information about the location of an object.

Some tracking schemes require more than one type of sensor for tracking. Each sensor can contribute to the overall goal of tracking. For example, accelerometers can be used to obtain information about relative motion but don't provide any information about an exact location. Additionally, many tracking systems that use accelerometers as their basis have a problem with error propagation. That is, any error, instead of being corrected, tends to become worse unless there is something in place to correct them. Because accelerometers are inexpensive, small, quick, and light, they are useful as tracking sensors when they are paired with additional sensors.

In addition to the sensors just described, other sensors are sometimes used as part of a tracking system. One of these is a GPS receiver. GPS provides a gross level of location information. Another sensor that can be used with tracking systems is a compass. A compass can be used to determine basic information about the orientation of the system. Further information about orientation can be discerned by using gyroscopes. Gyroscopes can determine data regarding relative orientation such as leaning, turning, and twisting.

2.3 COMPUTER VISION AND SOFTWARE FOR AR

2.3.1 Computer Vision

In AR, it is necessary to understand and be able to interpret the world around the user in terms of both semantics and 3D geometry. While the human brain is extremely good at understanding images, it still remains a very complicated job for computers. There is a complete branch of computer science, which is dedicated to understanding and interpreting images, called computer vision. Understanding semantics and geometry of the world around a user is a major task when developing AR applications. Traditionally, computer vision techniques used for understanding these two aspects are quite different. Semantics answers the "what?" question, for example, recognizing the object (a table being recognized as a table), or that there is a face in the image. Geometry answers the "where?" question and infers where the table or the face is located in the 3D world, and which way they are facing. Without geometry, AR content cannot be displayed at the right place and angle, which is essential to make it feel part of the physical world. There are different techniques for every domain. For example, computer vision methods that work for a face are quite different from those used for a table.

On the semantics side there has been much progress, all credits go to deep learning, which typically figures out what is in an image without worrying about its 3D geometry. As computer vision improves at understanding the real world around us, AR experiences would become more immersive, realistic, and exciting.

2.3.2 Key Approaches to AR Technologies

Marker Tracking: Augmented reality markers or short AR markers are visual cues that trigger the display of virtual information. Markers are normal images or small objects which are trained beforehand so that they can be recognized later in the camera stream. After a marker is recognized, its position, scale, and rotation are derived from visual cues and transferred to the virtual information.

Multiple Camera Infrared Tracking: In this method, an infrared projector is used to project markers onto the surface of the object. These markers are invisible to the human eye, but they could be identified by the infrared camera. This system consists of a conventional camera (scene camera) and an infrared camera (tracking camera). The viewpoint could be calculated, and the visible image is captured simultaneously when using such a camera system. Some 3D models are overlaid on top of a video image.

Simultaneous Localization and Mapping (SLAM): In computational geometry and robotics, simultaneous localization and mapping (SLAM) is the computational problem of constructing or updating a map of an unknown environment while simultaneously keeping track of an agent's location within it. While this initially appears to be a chicken-and-egg problem, there are several algorithms known for solving it, at least approximately, in tractable time for certain environments. Popular approximate solution methods include the particle filter, extended Kalman filter, covariance intersection, and GraphSLAM. SLAM algorithms are used in navigation, robotic mapping, and odometry (use of data from motion sensors to estimate change in position over time) for virtual reality or AR.

SLAM algorithms are tailored for the available resources and therefore not aimed at perfection but at operational compliance. Published approaches are employed in self-driving cars, unmanned aerial vehicles, autonomous underwater vehicles, planetary rovers, newer domestic robots, and even inside the human body.

The basic process involved in SLAM is presented in the following paragraph as a high-level view. When we first start an AR app using Google ARCore, Apple ARKit, or Microsoft Mixed Reality, the system doesn't know much about the environment. It starts processing data from various sources—mostly the camera. To improve accuracy, the device combines data from other useful sensors like the accelerometer and the gyroscope.

Based on these data, the algorithm has two aims:

1. Build a map of the environment
2. Locate the device within that environment

The SLAM system isn't a single specific algorithm or software, but rather is a set of algorithms that aim to solve simultaneous localization and mapping problem.

2.3.3 Software for AR

The popularity of AR applications has given rise to development of various tools. At present, developers who plan to get into AR have a huge variety of SDKs to select from. There are a variety of software tools involved with AR applications, whether it is the software running at the time the AR application is being used or the software used for the creation of content for an AR application. The good news for

AR developers is that very powerful tools are available for them that address the different aspects of application development.

The software involved with developing and using an AR application can be divided roughly into the following four categories:

- Software involved directly in the AR application
- Software used to create the AR application
- Software used to create the content for the AR application
- Other software related to AR

Another way to conceptualize the software components for AR applications is

- Low-level programming libraries (e.g., tracking software)
- Rendering and application building libraries
- Standalone applications (e.g., content building, complete AR authoring)
- Plug-in software for existing applications

2.3.3.1 AR Libraries

AR libraries are commercially available as well as freely download-able from various sources. Each library consists of a set of capabilities and a philosophy for developing AR applications. Hence, it is important to pick one's AR library wisely, taking into account the capabilities the library offers, the needs of the application, the target deployment platform, the level of support available, the development methodology, and the cost of the library.

Some of the questions to ask oneself when deciding on what AR library to choose: "What platform(s) am I developing for?"; "What sensors do I need to integrate?"; if planning to use computer vision for tracking, "What kinds of fiducial symbols do I need to use or do I need to use Natural Feature Tracking (NFT) or something else?" and "How many markers do I need to track simultaneously?"; "What kinds of graphics and audio are supported?"; "How much can I afford to invest in development time vs. cost to purchase a higher-end AR library?"; and so on.

There are a few AR development toolkits that need a mention here.

- Vuforia AR SDK, formerly Qualcomm's QCAR, is a Software Development Kit for creating AR applications for mobile devices.

- Wikitude SDK is an AR SDK for mobile platforms originated from the works on the Wikitude World Browser app by Wikitude GmbH.
- ARKit, an Apple SDK, currently designed exclusively for iOS 11+ app creation.
- ARCore, a Google SDK, currently designed exclusively for Android 8.0+ app creation.

AR libraries are becoming more and more powerful. Some of these are device-specific, while others are targeted at specific applications, and some of them are completely opensource. A detailed description of these libraries is available in the following chapter.

Although there are software tools for creating fully multimedia content in one package, more often, content creators use tools built and optimized for creating one type of content. At some point, authoring AR applications will be supported by easy-to-use tools that will create AR applications that can be used on a variety of hardware platforms. These tools will make it as easy to create AR applications as it is to make other types of software and will be suitable for use by designers, teachers, kids, and so forth. Unity is known to be the most famous and popular game engine around the world. In spite of the fact that it's typically utilized for creating PC games, it can very well be used for making AR applications with amazing results. As of 2018, approximately 50% of mobile and 60% of AR and virtual reality content have been created using this unity [7]. Approximately 90% of the AR/VR content has been created on emerging AR platforms, such as Microsoft HoloLens, and 90% of Samsung Gear VR content [9].

2.4 AR TECHNIQUES: MARKER-BASED AND MARKERLESS TRACKING

2.4.1 Marker-Based AR

Marker-based AR: AR brings digital information and virtual objects into 3D physical space. Simply put, it is a virtual representation of an extra media layer on top of a baseline media layer. One of the many cases in which we need to know "where the user is looking at" is called a marker-based technique. The baseline layer is a marker in this technique, and the media is displayed in 3D space when the user points his/her device at the marker. The smart device can recognize the marker as visual points such as logos, barcodes, images, and posters of any kind. A simple example of marker-based AR would be displaying a video of ancient culture at the museum when the user points the camera of the device to a small image located on the wall representing the information about the same old culture.

Marker recognition can be local or cloud-based, wherein the triggers or marker contents such as poster and images can be stored on the device (integrated into the compiled application), or the recognition happens on the server [15]. The difference between both the methods described above is that device-based identification is quick as images are available in local storage. In contrast, cloud-based recognition has latency issues because of the upload and download time of virtual content from the cloud. The most significant advantage of the marker-based technique is a quality experience and stable tracking of virtual information as the AR on the device does not shake as long as the marker is intact on the screen [10]. The applications equipped with marker-based AR are easy to use and don't require a long how-to-use manual. There are not many opensource platforms available to build marker-based AR, but Overly is a website that allows the individuals to upload the markers, overlaying content on the marker, and build an app for android, apple devices, or both with just a few clicks. The application, however, is paid and inefficient as you cannot redo the changes and ramifications in the application design. One has to go through all the publishing process again to make one modification (Figure 2.1).

Figure 2.1 Marker-based AR technique. (https://help.evolvear.io/viewing-ar-content/)

2.4.2 Markerless AR

This is the most challenging technique to implement in the real world because the user needs to design the virtual contents, feed in the application, and then provide a user with an opportunity to drag and drop the virtual materials in a real-world environment. In a marker-based AR approach, the user decides where to point the device in order to view the virtual contents, but in marker-less AR, the user needs to find a place to position the contents in a 3D space. The objects will appear to be "floating in the air" on the device screen with touchless interaction. In comparison with the marker-based technique, the marker-less technique provides the independence of choosing the place and object to the user and to create a virtual scenario with abstract elements to enhance decision-making. Imagine a mobile application for a furniture shop where all the types of furniture in the shop, such as chairs, tables, wardrobes, dining tables, etc., are virtually embedded in an application as objects. The users at their homes can use the app to place the objects in their background and judge the view: If it's looking good, color similarity, contrast, and various perspectives. The user can later conclude ordering only those items which get along with their house background, wallpapers, and other elements. Technology can enhance an individual's decision-making and can save a lot of time and effort of actually going to the shop and making choices.

2.4.3 G.P.S./Location-Based AR

The location-based AR technique makes use of real-world locations and coordinates to establish ties with the application. The trigger here is some unknown point in the real world where you can be the virtual objects popping up on the device screen as soon as the users reach that specific location. If you look at Google Map AR, the fancy glowing arrows for directions, no-horn zone, the speed limit is displayed as an extra layer of information on the user's screen when the application is in use. Let's discuss a few things about an accessible location-based game, "Pokémon Go." The game is a merger of location-based AR and marker-less AR techniques. The game has the world map integrated with the application where a character of choice represents the user. There are specific predecided points on the map which trigger the virtual objects (Pokémon's in the game) visible to the user. The game requires the location or G.P.S. to fetch map data from the cloud and trigger objects around the user's location. There is no marker adhered

Figure 2.2 A simplified AR merger of location and marker-less technique: Pokémon Go!. (https://ispr.info/2016/07/05/pokemon-enters-augmented-reality/)

to any Pokémon and is simply linked with the location; hence, the game resulted in being a merger of two AR techniques. After a successful AR implementation, several other games after Pokémon Go were published, such as Minecraft, Harry Potter Wizards, and others not as popular as Pokemon Go. Figure 2.2 is a snapshot of the game extracted from the original company website, Niantic!

2.5 AR DEVICES AND COMPONENTS

2.5.1 AR Devices

The AR ecosystem can be broken up into three classes based on devices suitable for developing AR apps:

1. Head-mounted systems, for example, Microsoft HoloLens (Figure 2.3)
2. Smart glasses like Google Glass and competing products
3. Handheld or mobile-based AR

2.5.1.1 Head-mounted Systems

Head-mounted systems or headsets are built to deliver highly immersive experiences that mix augmented and virtual reality environments. Head-mounted displays can be classified into

1. Video see-through HMDs
2. Optical see-through HMDs

Figure 2.3 Microsoft Hololens. https://www.microsoft.com/en-us/hololens/hardware

1. Video see-through systems: Video see-through systems display video feeds from cameras inside head-mounted devices. This is the standard method with which phones use AR. This can be of use when you need to experience something distantly: a robot which you send to fix a leak inside a chemical plant; a vacation destination that you're pondering about. This is also additionally helpful when you use an image enhancement system: a thermal imagery, night-vision gadgets, and so forth. The HTC Vive VR headset has an inbuilt camera which is frequently used for creating AR experiences on the device.

2. Optical see-through HMDs: An optical head-mounted display (OHMD) is a wearable gadget that has the ability of reflecting projected images allowing the user to see through it, similar to AR technology. Optical see-through systems combine computer-generated imagery with "through the glasses" image of the physical world, usually through a slanted semi-transparent mirror. If you are in a mission-critical application and you're concerned what would happen when you encounter a power failure, an optical see-through solution will help you to see something in that adverse situation. Microsoft's Hololens, Magic Leap One, and the Google Glass are recent examples of optical see-through smart glasses.

Figure 2.4 Smart glasses. (https://www.theburnin.com/wp-content/uploads/2019/09/Facebook-Smart-Glasses.jpg)

2.5.1.2 Smart Glasses

Smart glasses are low-power and lightweight wearables that provide first-person views. These glasses augment the user's vision. When wearing them, users can see their physical surrounding in the same way as in the case of traditional glasses. However, AR smart glasses like Google Glass superimpose virtual content to whatever the user sees. Vuzix Blade is another popular AR smart glass. It's a pair of AR smart glasses that float a screen in the upper right corner of one's vision. The company has partnered with Amazon to bring Alexa integration to the device, making the Blade the first pair of AR glasses to make use of Amazon's voice-based digital assistant (Figure 2.4).

2.5.1.3 Handheld or Mobile-Based AR

In handheld or mobile AR, all that is needed is a smartphone to have access to a host of AR experiences. A handheld display employs a small display that fits in a user's hand. The two main advantages of handheld AR are the portable nature of handheld devices and the ubiquitous nature of camera phones. Mobile apps on phone and tablet platforms are focused on AR overlays that leverage a combination of the device's processing power, camera lenses, and internet connectivity to provide an augmented experience. The rise of handheld AR is the tipping point for the technology being truly pervasive. AR libraries like ARKit, ARCore, and MRKit have enabled sophisticated computer vision algorithms to be available for anyone to use to build multiple AR apps (Figure 2.5).

Figure 2.5 Using handheld or mobile device for AR application. (https://www.intelivita.com/wp-content/uploads/2020/07/Google-Maps-AR-mode.png)

In addition to the methods and devices mentioned above, projectors can also be used to display AR contents. The projector can throw a virtual object on a projection screen and the viewer can interact with this virtual object. Projection surfaces can be any surface such as walls or glass panes [11].

Projected AR is a technology that directly overlays digitals projections onto the physical world. Unlike smartphones or wearable AR, projected AR does not typically require a device to mediate and project imagery. This creates the possibility of shared AR experiences and mixed reality experiences that integrate tightly with the environments in which they are installed. Projected AR systems make use of varied machine vision technology, often combining visible-light cameras with 3D sensing systems such as depth cameras. This allows the form of AR to use a technique called projection mapping, where the projected image is mapped onto physical objects, creating direct digital overlays. This allows for digital displays to appear on any surface or object.

2.5.2 Components

The three core components include the following:

1. Sensor(s) to determine the state of the physical world where the application is deployed
2. A processor to evaluate the sensor data, to implement the "laws of nature" and other rules of the virtual world, and to generate the signals required to drive the display
3. A display suitable for creating the impression that the virtual world and the real world is coexistent and to impinge on the participant's senses such that he or she senses the combination of the physical world and the virtual world. These components have been discussed at length in Section 2.2 of this chapter.

2.6 AR FOR IMPAIRED USERS

AR can play an important role in improving the lives of specially abled persons in numerous ways. A major problem faced by such persons is the lack of independence, mainly when handling day-to-day affairs. Going shopping, either for household daily needs or for leisure, is an activity of our daily life. Every family or individual needs to do this frequently. For the disabled, however, shopping is not always an activity they look forward to. It is very difficult for such people, who are unfortunately confined to wheelchairs, to move around the super-market or retail store and choose their products. The major problem here is that the shelves are very high, and disabled people cannot reach out to some products. This is because in many cases, disabled individuals are not thought of when it comes to designing the shopping experience. If this technology is widely adopted, not only will disabled people find that shopping is easier, but they will also enjoy it more because they can take part in it.

AR apps can help handicapped people shop with convenience by pointing their smartphone or tablet toward the desired product and getting the details of the product they want. The application can show all the updated information such as the price, stock, or the expiration date. Once the person has made a decision to purchase the product, it can be placed in a virtual cart. Then an employee of the store comes in to collect each item during the checkout. For those who are unable to move their hands, AR glasses can be of great help. Smart glasses allow you to take a look at some products without having to lift them with your hands. As soon as you look at an item through the AR glasses, you get detailed information about the product you want to purchase. One might say that AR does not allow to do anything new, and the same result can be achieved by doing online purchases. But let's not forget that the online purchase itself does not count as much as the experience of doing it. In addition, the use of AR will make people feel more satisfied and fulfilled, since it helps them to do their daily tasks without being dependent all the time on anyone.

People typically rely heavily on visual information when finding their way to unfamiliar locations. For individuals with reduced vision, there are a variety of navigational tools available to assist with this task if needed. One potential approach to assist with this task is to enhance visual information about the location and content of existing signage in the environment. With this aim, a prototype software application has been developed, which runs on a consumer head-mounted AR device, to assist visually impaired users with sign-reading. The sign-reading assistant identifies real-world text (e.g., signs and room numbers) on command, highlights the text location, converts it to high contrast AR lettering, and optionally reads the content aloud via text-to-speech. Participants with simulated visual impairment were asked to locate a particular office within a hallway, either with or without AR assistance (referred to as the AR group and control group, respectively). Subjective assessments indicated that participants in the AR group found the application helpful for this task, and an analysis of walking paths indicated that these participants took more direct routes compared to the control group [12].

MOBILE AUGMENTED REALITY (M.A.R.) FOR COVID-19 PUBLIC HEALTH MEASURE

3.1 SOCIAL DISTANCING FOR COVID-19

In 2020, the mighty world took a fall on humanity, trying to cope with an existential health crisis. The global outbreak was declared by WHO, a respiratory infection causing a virus, also referred to as the new and novel coronavirus (COVID-19). Due to the evolving capacity of humans with technology, the recent innovations and developments in technology can be directed to tackle the pandemic challenges and aid in forming safety measures for the people.

The government has advised specific health measures for the public to reduce the spread of COVID-19 disease in the community. The ordinary individual health measures include practicing good hygiene, such as washing hands frequently (after every 30 minutes) and using hand sanitizers to clean and disinfect your skin surfaces. Talking about public health measures, maintaining a 6-foot distance between two individuals is the best way to avoid virus transmission and reception. Wearing sanitary masks is, of course, the primary method to reduce infections, regardless of the location of an individual.

Being a touch-invoked kind of a disease, the governments across the world have introduced a safety public health measure called "social distancing," which will help prevent the spread of the virus. This chapter will address the procedures for how, when, and where the mobile augmented reality (M.A.R.) technology can enhance social distancing among incapable individuals. The latter half of this chapter will shed light on the system working and proposed methodology and provide a manual on using the mobile application.

There have been various settings suggested for communities such as child and youth, business and workplaces, cultural and religious gatherings, large group events, and local gatherings. We will emphasize the "social distancing" public safety guideline. Being a contagious person-to-person spreading disease, COVID-19 impacts people the most when they are in close contact with each other. Social distancing means staying at home as much as possible and avoiding crowded places, gatherings, and hanging-out in public attractions. This safety measure will work only if all the people of a particular geographic location willingly agree to adhere to it to help prevent the spread of the disease. It cannot be denied that, in order to protect others from COVID-19, each individual must make sure to protect themselves first. The other safety measures, excluding social distancing, which need a mention are self-isolation, sanitizing, and wearing a cloth-sanitary mask.

3.2 PUBLIC HEALTH SAFETY USING AR

The motivation for this chapter arose because of the lack of following the required precautions due to essential human needs and also because of inevitability of the disabled users to stand the technology and safety guidelines. It is important for normal human beings to follow the precautions and also to understand and assist the disabled users with the right technology to follow the safety guidelines and stay updated with them for avoiding further spread of the virus. Two aspects which need to be considered when using AR technology are usability (knowledge on how to use the tool) and flexibility (procedures to follow when and where).

3.2.1 Touchless Nature of AR

Augmented reality (AR) is a virtual technology that allows the user to use their body gestures to invoke the functionality of the application without physically touching it. This also being called touchless interaction reflects on user's emotions and satisfaction of their privacy. Due to this very advantage of AR, virtual technology has been deployed in various sectors and industries where computers are major. The research paper "Touch-less interactive AR game on vision-based wearable device" published in Springer emphasizes the pervasive games where the user interaction with the game is established using hand movement/ gestures. A total of 11 gestures were constructed for this particular game [13]. Following the technological trend for AR, the applications stretched from gaming to military, telecommunication, and healthcare areas. In 2018, AR was first tested in medical diagnosis and was worth an application to aid doctors in the surgery process. The doctors benefited from the 3D marker-based visualization of the patient's organs. They used the gesture control along with the involvement of medical equipment to avoid contamination in the surgeries (web source: http://avrlab.it/augmented-reality-and-myo-for-a-touchless-interaction-with-virtual-organs/) [14].

The primary objective of this chapter is to describe the applied science of AR technology for resolving COVID-19 medical crisis by proposing a digital mobile application for paired and impaired users to help follow the social distancing guideline efficiently. The pivotal elements such as the nature of AR, types of AR such as marker-based, marker-less, and superimposed [15] and real-time applications of the same have already been emphasized and discussed in the previous chapter.

The safety measure described in the section above puts forward simple definitions of social distancing, the importance of sanitizing, and other healthcare remedies. We all know that these rules are challenging and difficult to keep up with everywhere you go. One will get in contact with the other person for seeking help and answers to various questions accidentally or indirectly without considering the COVID-19 situation. Now, relating this very theory with the impaired users who are suffering from vision, walking, and speaking difficulties finds it uneasy to cope up with the public safety guidelines because of incapability.

Considering the scenario in a retail store, the disabled or elderly users find it challenging to reach the shelves, retain information on any item from the store, find the location of any item in the store, and know how to use the item and other information of relevance. To retrieve the information, impaired customers completely rely on the store associates and have to interact with them for more details. There is a dependency on other people for impaired users, and therefore, they are at a very high risk of catching the disease.

According to the Canadian Survey in disabilities report of 2018, the highlights mentioned that more than 6.2 million individuals or 47% of the disabled people are aged 75 years and over [16]. According to a report published by the World Health Organization on March 11, 2020, "The COVID-19 virus infects people of all ages. However, evidence to date suggests that two groups of people are at a higher risk of getting severe COVID-19 disease. These are older people and those with underlying medical conditions." There is a golden possibility that the problem mentioned above for impaired users can be efficiently resolved, by integrating AR technology into smart devices. The M.A.R. will be targeting the retail industry as an area of application to make the most out of it for following social distancing guidelines. Observing the recent COVID-19 pandemic statistics, several virus outbreaks occur in local restaurants, liquor shops, malls, and homecare centers. Regardless of the pandemic situation, retail/grocery stores are the locations that are going to be full of people all the time. Medical experts have advised to continue the social distancing policy and using hand sanitizers in crowded areas for the collective good. The application proposed in this chapter will be useful and reliable for all the customers in any retail store as the main objectives of the application will be to minimize the interaction, close contacts with store associates, and saving time. The intentions of the application will result in abiding safety guidelines for COVID-19 and create a passage for innovation and a source of entertainment in retail stores.

3.3 AR MODEL SIMULATION FOR SOCIAL DISTANCING

3.3.1 AR Model for Social Distancing Guideline

For every AR application to be delivered to the consumers, it needs to pass a few evaluation steps. The Technology Acceptance Model (T.A.M.) has been previously referred to by many research practitioners and widely accepted as an AR application evaluation method for user acceptance. The retail sector's importance is expected to rapidly increase its popularity because of technological advancements and research into usefulness, functionality, and acceptance. The T.A.M. is an evaluation model whose objective is to identify the basic requirements for AR application to be expected by the users [17]. The technology acceptance model is a predominant theory to examine user technological acceptance since 1986. The model incorporates the user's beliefs and attitude into consideration to adopt new inventions [18]. We will be discussing the TAM AR model in relation to our retail store concept and describe all the block details about the proposed model. The below-given figure is an extended version of T.A.M. External variables such as enjoyment, perceived benefits, and the previous researchers have introduced utilization cost before the year 2013. More of the few recent additions—enjoyment—confirmed its significance in future research [19].

By following the T.A.M. cycle of procedures, we will describe each block relating to our proposed concept of using AR application in retail stores, gyms, and public restaurants to abide by social distancing guidelines:

1. Information Quality: The quality of media such as image, audio, or video used as overlapping information or extra layer of information displayed to the users when they point their device to the barcodes and images placed on the walls of the establishment.
2. Perceived Benefits: The benefits are essential if seen from the user acceptance point of view. For example, the benefit of the application to the market, stakeholders, and consumers should be equal and not biased from one end.
3. Personal Innovation: The application should possess an open-source attitude and an innovative using the very technology within the context of the present research. The user, on the other hand, should show a willingness to try new products and services. The user's desire can be a combined expression of the application's simplicity, usefulness on all platforms, the user's intention to use, and data behavior.
4. Enjoyment: The AR application should act as a source of entertainment for categories of users such as kids, youths, and middle- and old-aged people. The application can have entertaining audio and video for kids and informative content leading to other links for the other age groups.
5. Cost of Use: While designing the application, the researchers are supposed to consider effort cost, privacy cost, and other nonmonetary efforts toward the sacrifices made for the sake of application.

Based on the literature that has been reviewed (8–11), the present research hypothesizes that enjoyment, possibility of personal innovation, perceived benefits, information quality, and utilization cost are the primary antecessors of users' perceived simplicity of use and perceived usefulness leading to attitude, intention to use, and data behavior as shown in the proposed Technology Acceptance Model (Figure 3.1).

3.3.2 Participant Demographics

After the idea was conceived, during the "Development and Validation of TAM for M.A.R. for COVID-19 Public Health Measure" study, a group of 64 paired individuals and 18 impaired in four different

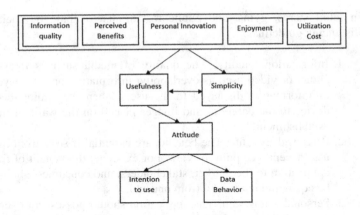

Figure 3.1 Proposed Technology Acceptance Model (latest development).

TABLE 3.1
Development and Validation of TAM for M.A.R.

| Age Group | Development and Validation of TAM for M.A.R. | | | | Testing Phase |
	Retail Stores/ Grocery Stores	Gym	Public Restaurants	Club House	Retail Stores/ Grocery Stores
Teenagers	5	2	8	0	24
20–40 yrs	15	7	9	4	18
41–60 yrs	9	3	7	3	34
Above 60 yrs	2	0	3	5	6
Males	10	6	12	4	35
Females	21	6	15	8	47
Impaired Users (blind, disability of the limbs, deaf and dumb)	7	2	7	2	14

application areas were surveyed, and in the testing phase, the same set of questions were put forth to a group of 82 participants in the first use case (retail only /grocery store) (Table 3.1).

It has been concluded from this study that the desire for high quality information about various items kept in the retail store was mostly preferred by the above 60 age group and impaired end users; enjoyable features and content (preferred by teenagers), perceived benefits (preferred by 41–60 yrs age group) and cost benefits (preferred by 41–60 yrs and above 60 yrs age group) as well as innovativeness are the key requirements of the end users. It is these five factors that decide the behavior, attitude, and acceptance of the technology by the end user [20].

3.4 SYSTEM METHODOLOGY

3.4.1 Proposed Methodology

The proposed methodology can be explained in two phases such as 3D content creation and AR content creation (Figure 3.2). The phases can be explained individually with highlighted points as follows:

- **3D Content Creation:** The real object is recorded through the digital camera of the mobile device as an image. The image needs to be deciphered and is forwarded to virtual engine such as Vuforia in this case. The image uploaded to the engine has to pass through the eligibility criteria of 200×200 in size and 8 bit for greyscale and 24 bit for RGB image. To add a target (image in this case) to the target manager in Vuforia, the type asked is single image, cuboid, cylinder, and 3D object. Typically, singe image approach is used, but if one wants to display a 3D model as overlaying information, the 3D object can be selected as the type for the target. After successfully adding the target to the manager, the features of the target are auto-updated, and the image is rated for "augmentable" out of 5 stars. The one with five-star rating is a strong marker.
- **AR Content Creation:** This stage is the final stage where the AR experience is displayed to the user on scanning the marker. The 2D/3D model is imposed as a marker and can be imported into Unity by simply downloading the database and importing it as an asset in the Unity's project view. The model adjustment is a touch-up step which makes sure that the marker is successfully scanned by the camera device without any system error and triggers the overlaying information if scan is positive.

3.4.2 Significance of the AR Application

The AR application will have its major contribution to the disability access needs in retail stores and will tackle the challenges by providing appropriate solutions to the needs. The application can also be

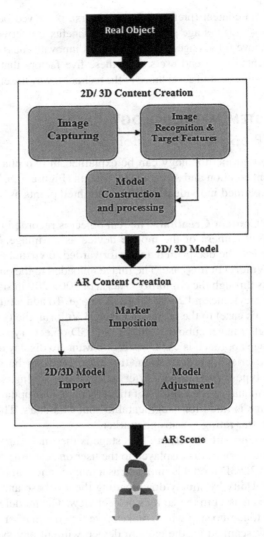

Figure 3.2 Proposed methodology for creating AR content and application.

deployed in public locations such as gyms, restaurants, clubhouses, camping grounds, and other local places where social distancing guidelines should be followed for public health safety [21,22]. Most of the retail stores across the world can usually occupy up to 300–600 people in the store at the same time, and hence there is a desperate need to take measures for people to abide by social distancing to help prevent the flow of the coronavirus. We targeted retail stores for using this proposed method, to address a few concerns, which can later be applied to other public places. The grocery store data can be analyzed for finding out the conclusions or answers to the questions. Every item in the sections of the stores such as grocery, pharmacy, seasonal, health and beauty, home décor, and others can be fed to the application to notify the location of the item in the store. This approach can result in time saving and minimal interaction with store associates. Usually, the people who are looking for certain products are not sure how to use it or simply don't understand the procedure of setting it up. The proposed research can deliver a solution for the problem as mentioned above, by adding a layer of information on the product image. The extra layer of data can be any media, including an image, instruction manual, price match options, and video sources. The extra layer of information can be mapped on a physical marker to provide details of the product to the users. The media modelling can be attained in 2D and 3D forms by applying core physics and coding concepts. The novel approach can ease the lengthy process of retrieving verbal information in the store and help achieve social distancing.

The proposed method can result in countless benefits to nonprofit organizations all over the world. Some of the organizations to benefit have been described below:

- From the North American perspective, the application has the potential to benefit such nonprofit organizations as Community Living Ontario; ideas can originate from Ministry of Children, Community and Social Services; and an entertainment aspect could be added to the application to encourage disabled kids and benefit such children care organizations as Empowered Kids Ontario, Autism Society, CanChild, and other relevant organizations.
- Healthcare homes and nonprofit organizations in India, such as Give India, Smile Foundation, Teach For India, and most importantly for senior citizens of India—Help Age India. There are many local N.G.O.s in all the states of India which are progressing positively toward the welfare of needy people. The knowledge about COVID-19, social distancing, and AR application can be conveyed in easy ways through these N.G.O.s in India.

3.5 USE OF UNITY, VUFORIA, AND C# FOR AR APPLICATION

3.5.1 Tools and Technology

AR is immensely popular in the 21st century because of its touchless nature and interactivity with the people. Companies use a globally trending tool such as Vuforia and ARCore to showcase a demo of their newly launched products to the market. Technical proficiency is not wide for this kind of technology and does mainly rely on developing IDEs such as Unity, Unreal, Visual studio, and other locally owned IDEs. One must possess great knowledge of C#, Unity, and Unreal engines to develop an AR application. Let's discuss a few renowned tools required to build an AR application.

3.5.2 Vuforia

Vuforia is the top-rated tool in the AR industry for developing AR applications. It is well known for its multi-feature performance, dual-platform support, and easy synchronization with Unity IDE. Vuforia has got its online database manager, where the user can store all of the marker's layered information. When it comes to textual data, it also supports text analysis and recognition. Engine recognition and tracking capabilities can be used on a variety of objects and images (Figure 3.3). The other important features of Vuforia are as follows:

1. Object targets: The targets are created by scanning the object and are a good option for toys and other products with consistent shape and rich surface details.
2. VuMarks: These are inbuilt markers that can encode certain data formats and can further be used for AR application tracking and identification purposes.
3. Vuforia fusion: Designed to sense the capabilities of underlying devices and merges them with Vuforia engine's features to provide the best AR experience on devices.
4. Vuforia Target Manager: The online database manager stores the images and updates the features of images for the augmentation process. The license key of the database is supposed to be added to the development engine (Unity works the best with Vuforia), and then the user can directly select the image uploaded to target images from Unity and begin to work with

Figure 3.3 Image recognition and target feature detection by Vuforia.

the AR application. The image shows the image features which the target manager (Vuforia online) identifies without additional efforts and enables connection with the IDE.

3.5.3 ARToolKit

ARtoolkit is an opensource software library under GPL (General public license) for building AR applications. The common problem in developing AR applications is tracking the user's viewpoint. The application should be aware of where the user is looking in the real world. To give a solution to this challenge, ARtoolkit makes use of computer vision algorithms to calculate the position of real camera and the position of physical markers in the real world. A few advantages of ARToolkit over other tools are as follows:

1. Simple black squares for tracking code
2. A free and opensource tool
3. Single-camera orientation
4. Great camera calibration
5. The toolkit can be downloaded for any platform from http://www.hitl.washington.edu/artoolkit/download/.

"ARtoolkit+" is a plugin that can be added in Unity development IDE for getting started with AR development with ARtoolkit+. The functional Unity schematic and a step-by-step tutorial for ARtoolkit+ with Unity is elaborated on the web source: http://www.arreverie. com/blogs/getting-started-with-artoolkit-unity-plugin/.

3.5.4 Google ARCore

Google has its platform for every popular opensource software such as Google Colab for Python-keras-Tensorflow and Firebase for android and owns a platform for building augmentation reality applications called ARCore. To integrate virtual content in the real world, the platform makes use of three core key features such as motion tracking, environmental understanding, and light estimation. We won't dive deep in all the features as Google ARCore won't be ideal for an AR experience in gyms, retail, restaurants, etc. for abiding social distancing. Not to mention the capability of ARCore to track the mobile device position as it moves and builds an understanding of the real world. To add to the quality list, the Google product can detect tables, chairs, and flat surfaces to estimate the average lighting nearby. The devices made to the "supported devices" for ARcore include android emulators, android for India, China, and Apple's iOS. The platform is opensource and can be quickly integrated as a plugin into Unity, Android IDE, Unreal, and iOS. Interested readers can learn more about the integration and tutorials at https://developers. google.com/ar.

3.5.5 Apple ARKit

Apple's leading platform for AR experience is promisingly delivered through ARkit for millions of users on iOS and iPads. The new ARkit 3.5 uses LiDAR Scanner and depth-sensing system on iPads to make realistic AR experiences for the users. The recent advancement to ARKit includes instant AR placement, people occlusion, and improved motion capture. Multiple face tracking, getting an estimate of the physical size of the object in an image, live collaborative sessions, simultaneous front and back camera, and machine learning algorithms deployed to detect planes in the environment even faster. To summarize the tools and technology section, the chronological order for a favorable tool to build an AR application would be Vuforia, Google ARcore, and ARtoolkit, and the list can have Apple ARKit at the end. To justify the chronology, Vuforia can deliver an optimal solution to all the issues and challenges faced while designing an AR application. The essential features of Vuforia such as motion tracking, Vuforia fusion, integration with IDEs, and target manager are sufficient requirements to deploy an application with efficient AR experience on all the platforms, i.e., Apple, android, and emulators.

3.6 HOW TO USE THE APPLICATION?

As already known, AR is a technology which has the capabilities of laying digital information layers on the real world and cover the virtual objects in a 3D space. The concept of an AR application sounds very futuristic and fascinating but is a challenge to reproduce one variation of the application for other areas. For example, an AR application created for retail stores won't give similar outcomes if deployed in the food industry chains such as restaurants and bars. The database, interface, overlaying information, and media should be updated according to relevant content to reproduce the application usage in other fields. The AR application proposed in this book will be developed using opensource software, and therefore, ethically, the application needs to be deployed as opensource under GPL for public use. We will be talking more about ethical issues, privacy, and security in regard to the application in the next chapter.

The mobile hardware configurations to run the AR application are as follows:

Mobile device: Apple device, any android device (Android 4.0 or higher), and android emulators

AR ready phone list:

- Google Pixel, X1
- Google Pixel 1,2,3
- Google Pixel 2XL, 3XL, 4, 4XL
- Motorola Z3, Z3 Play
- Motorola Moto G6, G6 Play
- L.G. G6, G7, ThinQ
- Samsung Series: Galaxy A70, A71, S6, S7, S7 Edge, S8, S8 Edge, S9, S9+
- OnePlus Series: 6, 6T, 7, 7T, 7 Pro, 8, 8 Pro
- Apple devices: iPhone 8, X, XR, X.S., 11, 11 Pro
- Apple devices: iPad Mini, Mini2, Pro, Pro (v2)

Besides mobile configurations, the application can be downloaded from Google Playstore for Android devices and Apple Store for iOS devices for no charge/in-app purchases.

Other functionalities the application might have or will be required by the user are given below:

- User sign up and login
- The application can remember the password for the user
- The user data will be encrypted using encryption standards
- Personal data will not be revealed to the public
- No cloud synchronization provided for security purposes
- Google sign-in not required
- The application may affect the battery life of the device as long as it is enabled.

4

EXPERIMENTAL ANALYSIS
AND RESULTS

4.1 MARKER-BASED AR TECHNIQUE

4.1.1 Introduction of Marker-Based Tracking and Marker Detection

Augmented reality (AR) is a technology used to present digital information in a real-world context. To do this, the system needs to know where the user is and the direction where he is looking. The user in turn is holding the camera-attached device, and the correct camera calibration is needed to render virtual objects in the right place. Marker-based tracking refers to calculating the location and orientation of the camera in real time. Previous researchers have developed a considerable number of tracking methods by using computer vision and photogrammetry techniques. A camera has always been an integral part of the AR environment, and visual tracking is of particular interest in the building methodology of AR. A marker is a unique sign which a computer or any workstation with processing capabilities can detect from a video image using pattern recognition, image processing and computer vision algorithms. The technique defines the scale and orientation of the camera once the marker is detected. This very approach is widely used in AR and is referred to as marker-based tracking [23]. Being utilized in visual tracking in AR and mixed reality, marker-based systems outclass the other feature-based systems on certain occasions. The technique of interest is widely used because it is easy to implement, and at the same time, open-source toolkits such as Google ARCore, ARToolKit, and ARKit, discussed in Chapter 3, are available for public use under General Public License (GPL). Markers also provide convenient coordinate frames and correct scale and orientation of the camera, as mentioned previously. In this section, we focus on a simple marker-based system and approaches that can be used for marker-based tracking and discuss a few points on marker detection and pose estimation.

Machine learning algorithms and vision detection techniques are best at determining "bright" elements than identifying "colorful" features. It is advised for a marker to be bright to get detected easily under all circumstances. This is all because of the auto white balance of the camera and the view angle. The system should be able to measure the pose of the camera with four random points forming a square and with the help of a marker. The most common types of marker detectors are black and white squares. The initial stages of marker detection start with locating the edges of the marker in the image and then circle an outline around it. The next step after the initial phase consists of confirming and deciphering the marker identity. Once the marker is detected, the pose is estimated from the information obtained via the marker. The final step-wise description for marker detection comprising all the steps involved in the procedure is given as follows:

1. Image preprocessing
   - Initial low-level processing
   - Line fitting
   - Edge detection
   - Corner detection
   - Circling the corners for final detection
2. Discarding outliers in the marker
   - Rejection and elimination of nonmarker elements
   - High acceptance of potential marker elements
3. Deciphering of markers
   - Template markers
   - Decoding data of markers
4. Calculation of the marker pose
   - Quick estimation of marker pose
   - Accurate pose calculation

There is a prologue step "Image acquisition," which takes place before preprocessing and is a separate process. This step provides the image for the marker detection process and nothing significant. The order of execution of these steps can differ according to the algorithm variation. It depends on the AR engine; actually, there are parameters to be given to the AR engine to trigger the marker detection. The system may accept the marker even without investigating in-depth and may also reject it at any stage if it thinks the candidate is not worthy. We will discuss all the marker-detection steps mentioned above in short, without digging in detail.

4.1.2 Image Preprocessing

The system needs to obtain a greyscale image with higher intensity before the actual detection of the marker. If the copy is obtained in a color format (RGB), the system converts it into a greyscale image. This conversion process is called "Low-level processing," which uses high-intensity images. The later phases consist of edge and corner detection, where the sharp corners and boundaries of the marker are identified. Marker detection techniques use an adaptive threshold method to process low-level procedures such as line cutting, edges, and border detection. The edge detection technique usually is not preferred by high-end application developers because it is tedious and time-consuming. Instead of edge detection, the subsampling approach is preferred as it has a predefined grid [24]. The systems use different line fitting, detection, and sorting approaches based on edge sorting and other methods described in the cited paper mentioned above. Real-time AR apps usually distort only the detected edges of the marker to speed up the system. In their research, the authors have stated that every acute minor error in the detected edges and corners significantly affects the pose estimation of the Camera [25]. The sources causing detection errors can be listed as pixel quantization error, motion blur, incorrect threshold value, and noise anomalies. The trivial errors have the potential to hit hard at the accuracy, and therefore, the systems are designed to optimize the edges' location after initial detection.

To club all the steps together, let's start with the threshold marker being detected by the marker. The system will make use of the greyscale image to identify boundaries with higher accuracy. Motion blur is estimated from exposed marker edges and then decompensated for higher efficiency. The diagonal corners occur in several directions, and then edge fitting stabilizes these. If analyzed with higher accuracy, the total number of edges and boundaries may take a tremendous amount of time processing nonmarker information.

4.1.3 Discarding Outliers in the Marker

The priority of every real-time AR application using the marker-based approach is to process the marker information as quickly as possible and not waste time scanning irrelevant markers. The app prefers using a fast acceptance/rejection approach to speed up the process of scanning markers. The marker having maximum pixels

Types of Marker for acceptance/ Rejection test

Figure 4.1 Categories of markers for acceptance/rejection task.

in a square block is visible or identified faster than the areas with less pixel population. The multicolored markers having two colors, say black and white, are bipolar and easy for detection. The bipolar markers have a fast acceptance/rejection criterion. 2D markers have several edges between black and white, and producing darker images is called "Vignetting."

As given in Figure 4.1, the two markers in the extreme corners are missing pixels in between the frame and object in the middle. All the markers being bipolar (black and white), it will still be challenging to identify the markers by some engines (Vuforia and Google ARCore should be good with this approach) from a reasonable distance from the camera. Therefore, markers with missing spaces are not advisable to use in real-time processing AR applications. The middle marker looks like a Q.R. barcode (just for an example), and if observed carefully, there is a dark edge in every pixel block of the image. In other words, the marker in the middle of the figure is a binary/bipolar marker with several strong edges. An application can gain rapid knowledge of marker detection if they are to scan robust markers with bold edges. The markers are generally preferred to have a square perspective, also called a quadrilateral view. A geometrical definition for a quadrilateral has four sides and four corners, and it doesn't necessarily look like a square. The even count of sides makes the calculation easy for the application, and thus, this technique is preferable for the fast acceptance/rejection criterion in an AR system.

4.1.4 Calculation of Marker-Pose

The position of any object in a 3D plane refers to coordinates in the location. The location is mathematically expressed in three parametrical forms, such as (x, y, z) and (α, β, γ) for orientation on the plane. The pose of a calibrated camera can be determined from four qualified points in the real-world plane. An AR system can estimate a marker's pose in 3D coordinates using four corners of the marker in image coordinates. The augmented view also depends on the type of camera that the device is holding. There are two types of cameras: A pinhole camera and a digital camera. For an ideal pinhole camera, a digital camera in (x, y, z) location can have an extended transformation on the world coordinates by superimposing the ideal image coordinates on the camera coordinates. The coordinates differ on both the planes and depend on the camera's physical characteristics such as focal length, sensor orientation, and size.

The camera transformation, which involves a marker present on location, goes through a matrix transformation step. The superimposed ideal image and camera coordinates are mathematically shown as two matrices multiplying to form a homogenous pattern and explain optical distortions. For the last step, pose calculation, let's assume that the optical distortions can be separated from the camera model. The points in both the coordinates, such as undistorted camera coordinate and in world coordinates, are multiplied in the form of matrices. A final image is located on the world coordinates to the viewers. There are minor detection errors spotted in the augmented view on the world coordinates, which doesn't impact the 3D view on the screen but needs to be resolved to improve its efficacy. The common issues usually observed are as follows:

1. The detection of "z" translation is not as effective as other translations.
2. The application can crash at any second if the device is moved too fast in between two points.
3. The pose of a marker seen from the front is more uncertain than the pose observed from an oblique angle (the side view and the top view are different).

4.1.5 How to Obtain Constant Tracking and Stability?

Many AR apps have the tendency to detect marker separately frame by frame. The system can be made efficient by keeping history on the marker appearance and tracking them simultaneously. The history preservation on the marker makes the deciphering process easy. There is no need to detect the marker in the first frame if it's decoded in the right way for the first time. The application can pose the marker over time and detect outliers if it keeps continuous track of the marker pose. As we have discussed in the above sections, pose estimation with a camera depends on the frame rates, and the pose estimation can vary if the frame rate is slow. One of the significant advantages of a continuous tracking system is that there is no confusion between the markers present and previous position as in every frame for pose estimation. It assumes that the marker is near its previous position because of the history metadata. The other smooth advantages of having an application with continuous tracking and tracking stability are given below as follows:

- The marker edges may shift due to the nonoptimal threshold and can lead to imprecision. These imprecisions appear as oscillations, and an application with continuous tracking can reduce the oscillations with "proper filtering."
- A marker with big cell sizes (fewer cells) can ease the detection from a reasonable distance. The detection process becomes complicated when there are abundant cells with the same size or smaller, i.e., large amounts of encoding data for the application. An application with continuous tracking ability could overcome this challenge by using super-resolution images.
- An AR app's stability can be enhanced by applying filters like Kalman filter (KF) and extended Kalman filter (EKF) for predicting camera movements.
- SCATT algorithm can be implemented for stable tracking and computing of pose estimation. A combination of SCATT and Kalman filter could be used for real-time monitoring.

4.2 ADVANTAGES AND DISADVANTAGES OF MARKER-BASED TRACKING

There are a lot of previous research siding either marker-based tracking or marker-less tracking depending on the area of application. The authors in their research article "Comparison of marker-based AR and marker-less AR: A case study on indoor decoration system" gave a brief comparison of both the techniques based on the aspects of methods, position accuracy, stability, and hardware support [26]. In this section, we are going to highlight a few advantages of marker-based tracking technique and state its importance in regard to COVID-19 safety measure, our topic of interest.

- The camera pose estimation, relative position, or relative angle depends on the markers for marker-based AR.
- AR software development toolkit (SDK) is necessary to build a marker-based application due to marker implementation and digital information to be displayed after detecting the marker.
- In comparison with the marker-less technique, marker-based apps have higher position accuracy.
- The influence factors for marker-based technique are brightness, SDKs, and markers.
- Marker-based applications are supported by both the platforms, desktop, and smart devices such as mobile phones, tablets, PDA, etc.

4.2.1 Advantages of Marker-Based Tracking for COVID-19 Safety Measures

- Given an example of deploying the marker-based application in the applications of retails, the camera-pose and relative angle would depend on the marker, and the customers have to go to a specific location to browse the location of items, item content, etc.
- We are considering Unity with C# and Vuforia, the SDKs for building an AR application for COVID-19 public health safety.
- Desktop platform would not be given priority to abide by social distancing rule as people won't be carrying their workstations in hands when they are walking in a crowd.

4.2.2 Disadvantages of Marker-Based Tracking

Every device and technique used with the device has a list of pros and cons. The authors in their article published by Springer explained the evolution of the technologies to make the shift possible from marker-based to marker-less AR technique [27]. Marker-based method for AR application has a few disadvantages for which there are no existing solutions so far. The problems are listed below:

- The AR experience disappears as soon as the user moves from the marker location. There has to be a constant interaction between the user device camera and the marker. The continuous marker detection could lead to unshakeable AR experience for the end-users.

- The biggest and under-rated disadvantage of the marker-based technique is keeping up with the functionality in light conditions. The light conditions can be extremely dark or extreme light exposure. The situation can also be linked with faulty and dysfunctional digital cameras on the devices.
- The scanning or detecting huge markers in bad weather conditions is tedious and cannot guarantee accurate results.
- The marker has to have bright edges and corners in the image to make the tracking more stable. Light colors in the image coordinates will make tracking vulnerable to failure.

4.3 AR APPLICATION DESIGN FOR SOCIAL DISTANCING MODEL

Before starting with the application design for our application intended for public health safety measures to follow social distancing guidelines, the goals of our AR application have to be determined for implementing and relating it with our design. The objectives of our AR system are as follows:

1. To challenge the incapability of impaired users to acquire knowledge about items, products, equipment in the areas of retails, food chains, and gyms.
2. To create a virtual environment in the real world for the users to have a wonderful mixed reality experience on their fingertips.
3. To integrate the core principles of AR to help the masses with their daily life struggle.
4. To achieve feats those are limited in the real social world.
5. To deploy the application as an open-source application with owners' permissions to motivate the students, practitioners, developers, and other creative minds to enhance their imagination and innovate new technology that would reshape the future.

4.3.1 Proposed System Architecture

The current section describes the system architecture for the marker-based system to be developed for COVID-19 safety measures. The block architecture consists of the following modules:

1. A Digital/Pinhole Camera
2. Image Operations Module (Image recognition, capturing, Processing)

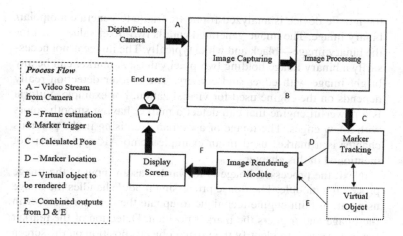

Figure 4.2 COVID-19 AR application system architecture.

3. Continuous Marker-Tracking Module
4. Image Rendering Module
5. AR Experience Display Screen
6. User Feedback

The proposed system architecture is given as a block diagram in Figure 4.2. The application design for COVID-19 AR application for public health safety will follow a similar system flow for high durability and tracking.

Each module has its own unique way of coping with the architecture and is explained in detail as follows:

A digital/pinhole camera:
The live real-world video stream is recorded through a camera and is fed to the image operations module in step 2. As shown in the architecture, the live feed is not stored in the local storage of the device but is fed to the module within the camera for image capturing.

4.3.1.1 Image Operations Module
This module consists of two interconnected sub-modules, such as image capturing, followed by image processing. The latter stage triggers an AR marker to provide the user with a virtual experience. To start with image capturing, the live video fed from the camera of

the mobile device is analyzed frame by frame to generate a bipolar/ binary image. The image generated entails two-pixel values used for the binary image—black and white typically. The image is not necessarily a binary image holding two-pixel values. A marker can be any bright image with edges and corners. The marker detection really depends on the engine used for virtualization. For example, Vuforia is a powerful engine that can detect a marker having smooth colors with a few edges. The image of a woman exercising in the previous chapter is a marker used in an example demo AR application not mentioned in this chapter.

Next, the processed image is forwarded as an input to the image processing module. The image processing module handles the image using an in-built engine technique to update the detector features in the image and triggers the marker position. Detection of the marker tag is necessary to identify the virtual object's position on the screen of the device. The same location/position is provided as an input to the tracking module.

4.3.1.2 Continuous Marker-Tracking Module

This module is the functioning backbone of the entire architecture that estimates the relative angle/position of the camera in the real world. To define "pose" in simple words, it is the 2D/3D location and orientation of an object. More information on the pose estimation is given in the last sub-section of this chapter's introduction section. The estimated pose after matrix multiplication is inputted to the image rendering module.

4.3.1.3 Image Rendering Module

This is a dual-input computation module that accepts two inputs and produces one output for the system. The two inputs are a pose obtained from the tracking module, and the other is a virtual object to be augmented. The rendering module is supposed to combine both the inputs, evaluate them for outliers/errors, compile for filtering and finally render the enlarged image on the mobile devices for the users. The rendering process comprising "n" steps is quick (takes a few microseconds). The engine ignores all the warnings at first, but not the critical errors. The errors related to code engine modifications need to be fixed before running the program. The best brute force approach is to restart the engine or application used to develop the program when such fatal errors occur.

4.3.2 Unity and Vuforia for AR Application

4.3.2.1 Unity Specifications and Features for AR

The aforementioned challenges can be resolved by using unity IDE along with Vuforia AR extension on the workstation to create posters and develop code theory. Unity 3D is a fully integrated environment with rich features that provide out-of-the-box functionality for creating 3D interactions. Unity is one platform for publishing applications on other platforms such as Xbox, PlayStation, web, iOS, and android. More details about how Unity works and other extreme features are discussed in this web source link [Appendix A]. The most beneficial features of unity development, which make it favorable for developing AR applications, are as follows:

- Complete toolset/SDK integration (Vuforia AR in this case)
- Intuitive workspace
- On-the-fly testing kit
- Feature editing
- Compilation of applications on major platforms such as android and iOS for mobile devices

Using other visual improvements of Unity such as shaders, textures, 3D packs for icons, buttons, and other graphical interface elements, the application can outclass the basic expectations of an AR application. Unity's AR feature support extends for ARCore, ARKit, Magic Leap, and HoloLens. The functionalities such as device tracking, plane tracking, anchors, point clouds, environment probes, light estimation, meshing, 2D image tracking, raycast, and session management can be used to create amazing AR experiences.

4.3.2.2 Vuforia Specifications and Features for AR

Vuforia AR is a powerful engine that supports cross-platform augmented and mixed reality with robust tracking and performance on a variety of hardware such as mobile devices, HUD, and HoloLens. Integrating Vuforia with Unity will allow the user to create streamline workflow using drag and drop techniques. Unity's asset store has ample of Vuforia AR+VR sample packages along with a demonstration of a few features. Vuforia supports both marker-based and marker-less tracking and has inbuilt algorithms to recognize the markers. The marker-less tracking relies on the device's GPS, accelerometer, hardware, and software. Being a powerful software, Vuforia

AR heavily relies on the hardware and software of the workstation/ device been used by the user.

The eligibility for mobile devices to run applications created using Unity-Vuforia SDK is as follows:

- Android O.S.: 4.1x and above
- iOS: 9 and above
- Windows: 2017+ version
- Applications for windows will run on 32 and 64 bit (auto-adjusted)

Vuforia's object scanner and calibration assistant are the two features which only run on Android 5 and above and iOS v9 and above. Vuforia links up with Google VR SDK, HoloLens and requires graphics API support from DirectX on Windows, OpenGL ES 2.0 on android and apple devices. More information on Vuforia platform configuration, tutorials, troubleshooting tips, and requirements is available on the web source https://docs.unity3d.com/2019.1/Documentation/Manual/vuforia_configuration.html [28,29].

Vuforia Integration with Unity

Unity supports Vuforia integration since version 2017 was released. To merge the SDK into Unity, download the latest release of Unity (v2019), go to the component dialog box in the initial steps and select Vuforia add-on. The download will start alongside unity setup, and the IDE will demand two more add-ons in order to get started with AR development with Unity. More tools required are given below:

- Java Development Kit (Latest JDK release)
- Android SDK tools
- Xcode
- iOS, universal windows support add-ons if compiling builds for the same platforms

The objectives should be aligned in the following order:

1. Setup Unity with C# editor and Vuforia integration to get started: Download and Install Unity's latest release and select other software requirements alongside installation such as JDK, iOS platform, universal windows platform and Vuforia AR for Unity. Start a new 3D project with a name, select the layout and window option from the top menu and check if you can see four panels labelled as Hierarchy, scene, inspector, and

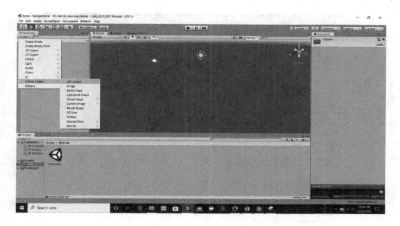

Figure 4.3   Unity and Vuforia setup.

project. As you can see in the Figure 4.3, the four groups are in default layout, and the perspective in the upper right corner of the scene can be controlled according to the user. The scene is set with the main camera and directional light at first. For importing Vuforia scripts, go to file → build settings → player → X.R. settings → check virtual Reality & Vuforia AR support. The scripts will be downloaded automatically upon checking the support boxes.

The next step will be to delete the main camera and directional light and add AR Camera from Vuforia Engine. After discarding the two elements, add an image from the same Vuforia Engine.

2. Convert image into trackable data using Vuforia: After setting up the Unity scene elements, Vuforia comes into play to add the overlaying image (to be augmented) on the image object we created in the above step.

Go to Vuforia developer website → login with user credentials → License Manager → Copy the license key in the Vuforia cameras Configuration license field (Figure 4.4).

After copying the license key into Vuforia configuration, go to the target manager on the Vuforia developer website and create a database (if a new user), and add a new target for feature identification. Vuforia updates the features of the marker automatically using a by default in-built algorithm. The image given below shows the marker added to the Vuforia database.

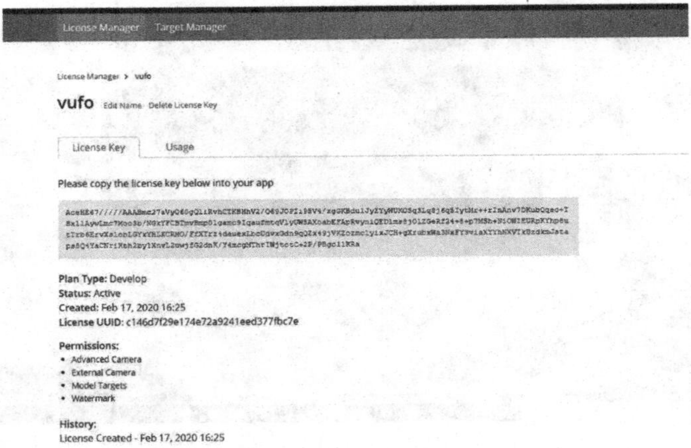

Figure 4.4   License manager setup in Vuforia.

Figure 4.5   Q.R. code AR features.

Once the marker is added, download the database and import it into Unity as asset import (Figure 4.5).

As you can see in Figure 4.6, the marker uploaded to Vuforia has status as active and five-star rating for "augmentable" feature. Now we go ahead and download the database to upload it to Unity.

Figure 4.6 Vuforia object inspector features.

Procedure to import the downloaded package:

Go to the "Project" window in unity → Import package → custom package → Select the downloaded package from the local storage.

After the successful import, go to "AR Camera," select the database from the "Image target behavior" from the inspector window in the layout and the image target as the marker uploaded to Vuforia previously.

3. Add overlays to trigger images: The media which is to be displayed to the user (the overlaying info) upon scanning the marker is to be added as an image/video. We go with the image for now, and one can simply import an image into unity asset space and specify the type as "Sprite or 2D Sprite." Once the image is specified as a sprite, drag the image to the marker image target. Moving forward, create a new GameObject by navigating to Hierarchy section "+" → empty child → GameObject and rename it as "finder" or whatever you like. Make sure to use the same name everywhere in the file.

The new overlaying image used in the AR application sample to show in this chapter is given in Figure 4.7. The image simply represents the quality content difference between the types of bread and nutrition values. We are giving an example of using AR in the grocery industry for avoiding interaction and abiding with social distancing guidelines.

Carbohydrate content of different bread varieties (per 100 g)*

	Fibre	Sugar	Total Carbohydrate
Rye	6 g	3.9 g	48 g
Focaccia	1.8 g	1.8 g	36 g
Sourdough	2.4	2.6	56
White	2.7	5	49
Wholegrain	7.9	2.4	30.8

*Estimates only, variation will be observed between brands.

Figure 4.7   Overlaying information on the marker.

4. Modify C# scripts using C# editor (Visual Studio, sublime, etc.): To modify the C# script, search for "DefaultTrackable EventHandler.cs" in the project sections search bar. Open the file with visual studio or any editor of your choice and add a few lines of code below.
   - Add a public GameObject in the protected member variables, which will remain static throughout the program.
   - Add finder as SetActive(true) and SetActive(False) in onTrackingFound() and onTrackingLost() to tell the application to disable the Finder game object is the target is found and to enable is caught.
   - The part modified from both the points above is highlighted "bold" in the code below.

```
using UnityEngine;
using Vuforia;

/// <summary>
/// A custom handler that implements the
ITrackableEventHandler interface.
///
/// Changes made to this file could be overwritten
when upgrading the Vuforia version.
/// When implementing custom event handler
behavior, consider inheriting from this class
instead.
/// </summary>
public class DefaultTrackableEventHandler :
MonoBehaviour, ITrackableEventHandler
{
    #region PROTECTED_MEMBER_VARIABLES
```

```csharp
    protected TrackableBehaviour
mTrackableBehaviour;
    protected TrackableBehaviour.Status
m_PreviousStatus;
    protected TrackableBehaviour.Status
m_NewStatus;
    public GameObject Finder;

    #endregion // PROTECTED_MEMBER_VARIABLES

    #region UNITY_MONOBEHAVIOUR_METHODS

    protected virtual void Start()
    {
        mTrackableBehaviour = GetComponent<Trackabl
eBehaviour>();
        if (mTrackableBehaviour)
            mTrackableBehaviour.RegisterTrackableEv
entHandler(this);
    }

    protected virtual void OnDestroy()
    {
        if (mTrackableBehaviour)
            mTrackableBehaviour.UnregisterTrackable
EventHandler(this);
    }

    #endregion // UNITY_MONOBEHAVIOUR_METHODS

    #region PUBLIC_METHODS

    /// <summary>
    ///     Implementation of the
ITrackableEventHandler function called when the
    ///     tracking state changes.
    /// </summary>
    public void OnTrackableStateChanged(
        TrackableBehaviour.Status previousStatus,
        TrackableBehaviour.Status newStatus)
    {
        m_PreviousStatus = previousStatus;
        m_NewStatus = newStatus;

        Debug.Log("Trackable " +
mTrackableBehaviour.TrackableName +
```

```
                " " + mTrackableBehaviour.
CurrentStatus +
                " -- " + mTrackableBehaviour.
CurrentStatusInfo);

        if (newStatus == TrackableBehaviour.Status.
DETECTED ||
            newStatus == TrackableBehaviour.Status.
TRACKED ||
            newStatus == TrackableBehaviour.Status.
EXTENDED_TRACKED)
        {
            OnTrackingFound();
        }
        else if (previousStatus ==
TrackableBehaviour.Status.TRACKED &&
            newStatus == TrackableBehaviour.
Status.NO_POSE)
        {
            OnTrackingLost();
        }
        else
        {
            // For combo of previousStatus=UNKNOWN
+ newStatus=UNKNOWN|NOT_FOUND
            // Vuforia is starting, but tracking
has not been lost or found yet
            // Call OnTrackingLost() to hide the
augmentations
            OnTrackingLost();
        }
    }

    #endregion // PUBLIC_METHODS

    #region PROTECTED_METHODS

    protected virtual void OnTrackingFound()
    {
        if (mTrackableBehaviour)
        {
            var rendererComponents =
mTrackableBehaviour.GetComponentsInChildren<Renderer
>(true);
```

```
            var colliderComponents =
mTrackableBehaviour.GetComponentsInChildren<Collide
r>(true);
            var canvasComponents =
mTrackableBehaviour.GetComponentsInChildren<Canvas>
(true);

        // Enable rendering:
        foreach (var component in
rendererComponents)
            component.enabled = true;

        // Enable colliders:
        foreach (var component in
colliderComponents)
            component.enabled = true;

        // Enable canvas':
        foreach (var component in
canvasComponents)
            component.enabled = true;
    }
    Finder.SetActive(False);
    }


    protected virtual void OnTrackingLost()
    {
        if (mTrackableBehaviour)
        {
            var rendererComponents =
mTrackableBehaviour.GetComponentsInChildren<Rendere
r>(true);
            var colliderComponents =
mTrackableBehaviour.GetComponentsInChildren<Collide
r>(true);
            var canvasComponents =
mTrackableBehaviour.GetComponentsInChildren<Canvas>
(true);

        // Disable rendering:
        foreach (var component in
rendererComponents)
            component.enabled = false;
```

```
            // Disable colliders:
            foreach (var component in
colliderComponents)
                    component.enabled = false;

            // Disable canvas':
            foreach (var component in
canvasComponents)
                    component.enabled = false;
        }
      Finder.SetActive (true);
      }

      #endregion // PROTECTED_METHODS
}
```

5. Compile the AR application for platforms—Android, iOS, and Windows.

If there is a webcam present on the development workstation, the developers can skip to the part where they save the current scene and compile the application for the platform that they wish to generate the application. The process mentioned in the next few lines is to be followed to create a successful build for mobile devices.

- Save the scene by giving it a name.
- Go to build and select "Android" or "iOS" (your choice)
- The system will check for JDK and android SDK/iOS development kit before switching platforms.
- If all the requirements are met, go to "build" and generate a build for your chosen platform.
- Save the apk/ios file in the local storage.
- Deploy the application on Cloud for user access

4.4 APPLICATION DEPLOYMENT AREAS FOR COVID-19 HEALTH MEASURES

The AR application development and working shown above is a use case of the application in "retail stores." The application merely shows the comparison of bread contents when been scanned by the application. The same technique can be extending to other examples such as "How to use the item" video when the user scans a product. Every shelf has a price tag at the bottom, and the compact Q.R. marker can be concatenated with the price tag showing a video /image/ reviews and other alternatives to the item. For example, many customers don't know how to use a "barbeque" and have doubts about purchasing an expensive item without much knowledge. The video implementation of the exact same brand and specifications in a video would save the customer's efforts and time of googling about the content, how it works, and reviews about the product.

4.4.1 Application Integration

There must be a question arising in the reader's mind asking "if they have to sacrifice their device's storage for the exclusive AR application for each store they visit"? The answer to this question is no; one doesn't have to download a hundred applications for hundred stores they visit. The piece of code containing AR dependencies can be embedded in the existing store applications. Walmarts across America and North America have their mobile application on all platforms, and they already use a barcode scanner in their application. Embedding a Q.R. scanner in the application or linking it with the barcode scanner won't be much of a tedious task for Walmart. In India, D-mart is a popular retail chain having thousands of customers every day regardless of the location. D-marts can implement the same technology in their application and serve the population with a new learning scheme.

4.4.2 Other Application Areas for Application Deployment

1. Public Gyms and Fitness Centers: Another use case example of the proposed AR application would be applicable in fitness centers and local gyms. The app won't be a replacement for a physical trainer, but it can tell the people how to use the equipment. Imagine you are a cardio person, and you see a set of kettlebells in front of you. You have no idea what to do with those, but you just know that it is an excellent cardio exercise. It would be straightforward to scan the marker on the shelf where they reside and to know more about how and what type of activities to do using kettlebells instead of going to google for a solution. The image shown on the corner right in Figure 4.8 shows an example of AR examples where the categories have been organized, such as kettlebells, dumbbells, bench press, and chest machines. The AR application is planned and is currently under development.

Figure 4.8   Other application areas for marker-based approach.

2. Cleaning and Janitorial Services: The marker-based approach has been used by building services and janitorial service providers in North America to make form-filling easy and avoid the load of work. As you can see in the figure above, a Q.R. marker is fitted on the washroom labels (first two images) and triggers the browser page when scanned by the camera of the mobile device. Scandinavian Building services (https://scandinavian. ca/) [30], a leader in Canadian retail industry and commercial cleaning, which provides services to Walmart in Thunder Bay, Ontario, Canada, have implemented the marker-based approach for the cleaners to avoid getting in the washrooms in the unprecedented times of COVID-19 and to abide by social distancing guidelines. The webpage for form-filling pops up on the Safari browser on an iPhone device when the cleaners scan the image using a camera. There are other areas in which the application can be used such as museums, art exhibitions, and public attractions where there can be more knowledge shared about the artifacts, products, and locations, and people maintain a safe distance from each other. The technology can be majorly utilized to its peak in the field of military and telecommunication where the ammunitions can be visualized as 3D models with the details to the buyers, users without the actual physical product being present at the location.

Following the current and previous chapters, the next short chapter will summarize the concepts and theories explained in this book and highlight all the essential points in the technical discussion.

5

MITIGATION REMEDIES AND RECOMMENDATIONS

5.1 MITIGATION REMEDIES AND RECOMMENDATIONS

The general information for COVID-19 prevention will always be to wash hands, sanitize the belonging area, and frequent cleaning of the residing location. High-touch areas such as light switches, doorknobs, laundry machine control panels, keyboards, and remote controls should be sanitized/cleaned and kept free from clutter.

Other mitigation remedies followed by most of the countries across the world are as follows:

1. Public and Private Gatherings: There has been a restriction on the maximum number of people present at one location that should not exceed the advised count. The limits may vary from country to country, depending on the number of COVID-19 cases in the respective region. Following the public gathering restriction, the public outings, the number of people present in the store at the moment, curbside pick-up, and family reunions are also impacted during the scare of a pandemic.

2. Food Handling: The risk of falling sick due to food ingestion in the COVID-19 season is pretty low. There are cases where the infected people without being aware of the infection walk in the food stores and might touch the packaging accidentally. Following social distancing guidelines in the local and retail food stores is the first follow-up, and a 2 m distance between people in checkout lanes is advised.

3. Self-Isolation in Hotels: The individuals out of their home county or home location feel that they are exposed to the disease or have come in close contact with someone carrying the infection. They can choose to self-isolate themselves in a hotel room on the following conditions:

- The individual should not leave the room once checked-in the hotel room.
- Regular housekeeping services should not be provided to the self-isolating individuals.
- Access to a swimming pool, sauna, gym, and bar should be prohibited to the individual.
- The room's high-touch areas should be cleaned or disinfected at once after the guest leaves the room.

4. The other impacted areas or the fields that need attention to help prevent the spread of COVID-19 are public households, transportation, apartments and buildings, shelters and food banks, and group livings that have the same restrictions and mitigation recommendations applicable.

5.2 EXISTING APPLICATIONS FOR SOCIAL DISTANCING

5.2.1 Google SODAR

To fight COVID-19, Google developed its own version of augmented reality (AR) application which will help follow social distancing effectively, an application called "Google SODAR." SODAR stands for social distancing using AR, which creates a 6 feet or 2 m radius ring around the mobile screen. The ring is represented on the screen as a form of visualization and to determine how far 6 feet is supposed to be from one person's angle. The technology is supported by all android and iOS phones capable of running AR on their chips. The application is not available as a standalone application which needs to be installed and can readily be used by simply visiting the website. The application can only be opened on chrome browser on supported phones.

5.2.2 Aarogya Setu

Aarogya Setu is an Indian contact tracing app which is created by India's Ministry of Electronics and Information Technology in response to the COVID-19 crisis and other measures. The application is not based on AR concepts and principles but is worth mentioning in this section because of COVID-19 context. The primary objective of the application is to detect and notify a positive COVID-19 case in a person proximity in form of alerts/notifications. The purpose is to make the person aware of a case in his/her proximity so that necessary steps could be taken in order to stay safe. The application was released in 11 languages and gathered personal information for registration and sign-up which led to numbers of privacy and security questions. The application mostly conducts "self-assessment" questionnaire which prompts the user to answer the health questions and collect the device location and upload to the server. The application monitors the users' data every 15 minutes and declares the flags such as orange, yellow, and green as user's health indication. There are a lot of privacy concerns which arose due to the data collection and the applications being opensource. There have been misconducts and access to the data collected by the application, and the details are available at https://www.lexology.com/library/detail. aspx?g=f54419a1-4823-404c-92f3-c5e4f193b733.

5.2.3 1point5

1point5 is an application deployed for public for educational purpose and not for monitoring. The application is not a COVID-19 assessment or medical application, but it educates people on social distancing and other public health measures. The novel feature of 1point5 application is that it alerts the user when someone is approaching your 1.5 m perimeter. Other features of 1point5 are device-based exclusion, distance range customizations up to 2 m, and facility to run the app on small devices such as smartwatches, phones, and mini-tablets. The app is available on Google play and compatible with almost all the devices.

5.2.4 DROR

This year has witnessed the launch of several tools and applications developed to tackle COVID-19 problems. DROR, a safety application developed in Delhi, India is a Bluetooth-enabled app which serves the purpose of social distancing tracking. DROR was deployed in action since August 2018, initially developed for women safety but later evolved with the rapid increase of COVID-19 cases worldwide. The application maintains the social distancing score for the users by exchanging short-distance Bluetooth signals between phones to detect others' use of the app who are in nearby proximity. The app is constructed on the pillars of machine learning and artificial intelligence algorithms and notifies the user if he or she gets close to a huge crowd / busy place. The only limitation of the application is Bluetooth and not internet connectivity. The Bluetooth should be enabled for the application to compute the scores.

5.2.5 Canada COVID-19: COVID Alert

Canada's digital contact tracing application was developed and launched on July 31, 2020. As countries all over the world emphasized on utilizing digital technology to prevent the spread of the novel coronavirus, Canada built its own digital contact tracing mobile application with the features same of Aarogya Setu of India. Just like all other applications described in this section, COVID alert notifies the person using the application about the COVID-19-positive cases in the proximity of the same person. The application also has assessments and news update features to provide to the users. There have been millions of downloads for this app since November, and the number seems to still increase. The application is available on Google Play Store and Apple's Appstore.

5.3 RECOMMENDATIONS OVER EXISTING APPLICATIONS

To recommend the purposed AR application over all the abovementioned COVID-19 existing applications, our application is an exclusive use case for abiding social distancing guidelines in retail stores amidst COVID-19. To discard the temporary COVID-19 limitation, we have added features to the application, to serve the purpose of an educational tool for the users. The specific reasons for selecting our application over others have been mentioned in the above sections and in Chapter 4 of this book. Google SODAR, COVID alerts, and other tools could be used to abide social distancing in all use cases/applications, but they all are uni-featured or single-featured applications and perform one single activity at all the times. The AR application which we have proposed has multiple features such as display of information about the product when scanning the price marker, comparison with other products of the same race, videos on how and where to use the product, and other educator features for entertainment. The next subsection "Technical Discussion" will discuss the shortlisted techniques, use cases, and others to build a quality AR application.

5.4 TECHNICAL DISCUSSION

The technical discussion for the topic of interest will emphasize on enhancing the AR application and address a few concerns such as "how to improve the relation of the technique with the outside world," "what are the practical experiences in AR development," and technology acceptance and adoption. The best rendering technique to enhance human perception depends on the purpose of the application. There is a constant interaction between real and virtual objects in an AR system, which profoundly affects the user experience. The interaction between the real and virtual objects may result in several occlusions, collisions, and minor fluctuations, which we will discuss in Section 6.1, labeled as issues and challenges in AR application development. In this section, we will elaborate on how to improve visual perception and deal with AR systems' practical experiences. Visual perceptions steer human attention in several ways and visualize the information in the correct environment. Any AR system can attract human attention with nonphotorealistic rendering, also called NPR method for visual perception. The overlaying information or the augmentation in this method is bright and distinctive to the user. The photorealistic rendering (P.R.), in this case, is the opposite of the NPR method, and both the objects, i.e., virtual and real, are indistinguishable from each other. The PR method is highly used in applications where high-quality visualization is required. By default, the method applies to the proposed social distancing-intended AR application, the NPR method, where the markers are usually bipolar/monochromatic, and the augmentation is bright and illuminated. In this case, the augmentation can give instructions and draw attention from the users for the next steps. The application is not used in the areas where high-quality visualization is required and will mostly be used for the general public visiting places quite frequently. Therefore, photorealistic perception is not much of necessity, and the nonphotorealistic perception method would be suitable in all the conditions. There might be circumstances when the AR application renders virtual objects on top of a real-world image without processing the aspects of the view. The result is quite unrealistic where the virtual objects float in the air with marker visible from edges and drawing unwanted attention. The virtual engine possesses features such as shadows and lights to feature to the augmented object where the light adjustment, according to the real-world light, will decrease the brightness of the virtual object and make it look naturally illuminated with the real world. The virtual shadows and light will touch up the perceived realism. Adding more to the enhancing user experience, the rendered virtual objects should have a similar appearance as of the captured image to achieve systematic and symmetrical imagery. Following up with the illumination characteristics, the application needs to be equipped with motion-blur and focus capabilities. By the word motion-blur, we understand that the virtual objects need to stay static and rendered on the screen even though the user is slightly shaking the device while scanning the marker. The future scope of the proposed application may include capabilities such as illumination, focus, motion-blur, and maybe even rendering to the real-world noise and distortions. An application having these features will be aspiring to achieve ultimate realism. The texture generation option for the markers by the virtual engine is a challenging procedure and would hide the markers' qualities to the users on the device. The next sections of this chapter will shed light on challenges faced in the AR application development and end the last chapter of this book with conclusion, acknowledgement, and bibliography content.

5.5 AR SOLUTION TO THE PUBLIC HEALTH COVID-19 MITIGATION REMEDIES

The AR development approaches elaborated in the previous chapters are about the two prime techniques such as marker-based and markerless AR We won't be discussing the details about these techniques in this section because we are going to focus on building a fair AR solution for the mitigation remedies mentioned above.

Being the preferred choice for constructing the proposed AR application, the marker-based approach can also be implemented to support the mitigation remedies recommended by the government. There can be a hybrid approach undertaken for developing a solution for food handling and public gatherings. The markers can be imposed on the food shelves or the food containers in a public gathering/function involving lunch or dinner plans, which will enhance the decision-making of the people attending the gathering. The people who are not sure of how the food tastes or the nutritive contents of the food can benefit from the application. They don't need to ask the caterers and can also have the information in the form of AR experience on one scan. For other mitigation recommendations such as gyms, swimming pools, and restaurants, the marker-based approach can be used to demonstrate the "how to use" manual in the form of a video or instructional image. In gyms, the markers can be placed on the equipment's shelves to explain the work-out procedure. The motive is to let people know the significance of the work-out, diet sheet for the week, impacts on the body, and other relevant stuff regardless of the trainer being physically present on the floor. The remedies are not only limited to the coronavirus precautionary measures and should be adopted permanently for the better good of humanity. Let me ask the audience, "Does sanitizing or disinfecting the area you've been or the things you've used to do any harm to you?" The answer to this question is mostly going to be "No" because it's a win-win situation for the people where they get a gesture of being safe and make the world a better place for the next generation. After struggling and surviving the cruel pandemic of 2020 together, people can only imagine the next pandemic and prepare a contingency for it. The contingency can be a merger of digital technology and innovation. Not only AR, but researchers have been using IoT and machine learning platforms to develop applications for keeping people safe daily. The research and innovation have made it possible for a person to know how many calories he or she has burnt after walking for 1 minute. The contact tracing applications such as "COVID Alert," introduced by the Canadian government, will notify the users if someone nearby their location or in their locality has been tested positive. The Indian government initiated the same approach in April 2020; the contact tracing application for India was launched as an opensource application with other unique features such as syndromic mapping and self-assessment methods. The mobile app was deployed to connect essential health services for the people by the people to fight the coronavirus crisis.

5.6 MERGER OF DIGITAL TECHNOLOGIES FOR COVID-19

The medical practitioners are working tirelessly to help prevent the new coronavirus outbreaks on local levels and providing efficient treatment to the already affected. During this vulnerable phase, science researchers and technology experts worldwide are collaborating to make use of cutting-edge technology to aid surveillance systems for better monitoring of people, decision-making for the infected, and periodical checks of those exposed to the virus. Apart from using Big Data, AI, and machine learning as vitals, scientists are shifting focus on the Internet of Things (IoT), a combination of IoT and cloud and IoT-related domains to mitigate and monitor COVID-19-exposed individuals. To recover the tremendous overall loss, the healthcare units in China are making use of "connected thermometers" to monitor real-time body temperatures of the hospital staff and patients and for finding counter measures [30]. Due to several known advantages of wearable computing, IoT experts have recommended using smart health bands to keep track of body temperature and using the consistent series data to forecast future temperature and detect drop or rise indicating medical attention need. Vancouver hospitals have deployed Visionstate's "IoT button" in few hospitals that report the lack of cleanliness, sanitizing, and maintenance issues in the staff or patient cubical on a press to the on-site manager [31]. The popular IoT devices such as Google Home, Amazon Echo, Tracking Fitbit, smart lights, smart toothbrush, and smart doorbells which connect to internet, provide services, upload data on cloud, and update the users have several limitations [32]. Talking about AR applications, more than half of the applications revolve around smartphones for AR technology. Newer inventions during COVID-19 pandemic have led to Google SODAR and 1point5 for abiding by social distancing rules. The case study of AR application mentioned in this book is a retail store use case which could be extended to other areas such as fitness centers and exhibitions where people tend to interact with others to acquire information or knowledge about some equipment (for gyms) or product in art galleries and exhibitions. The AR application can be developed as a merger of technologies such as AR, AI, ML, and VR. Machine learning algorithms could be used to predict the user's next move in order to avoid social distancing while walking (Google SODAR's extension), AI techniques could help the user for efficient decision-making, and inclusion of VR might just immerse the user in a 3D virtual environment. Nevertheless, while there are attractive features of the emerging AR technology, ethical practices and GDPR rules should be designed to monitor the impact of technology on humans. Trustworthiness, fairness, confidentiality, and security should be implemented in any smart device/application to protect user data and achieve powerful group dynamics. To follow the ethical framework and humane design guide concepts (described in Center for Humane design), the components embedded in the application development cycle are given below as follows:

1. Emotional: To secure the human emotions of fear and worry about the personal information, there is no personal information such as DOB, place, pin code, bank details required for installing, and signing up to access the information. Name and surname with one unique identifier such as mobile number or email ID are sufficient for login and signup.

2. Attention: The application does not hold the feature of infinite scroll and constant streaming so as to acquire the constant attention of the user.

3. Sense making: The commercial advertisements to the free using customers can divert user's attention and break the contact from the application. There should be a sensible flow of information in the stream of the application.

4. Decision-Making: This is the primary concept for our proposed application where the user will only scan and fidget with the app when he or she knows what item to browse/access/shop or gather knowledge about. Efficient decision-making will lead directly to the usage of the application.

5. Group Dynamics: Several users accessing the app could connect on a virtual platform on a specific page of the application and share the information obtained for various products.

6

SUMMARY

6.1 ISSUES AND CHALLENGES IN AR APPLICATION DEVELOPMENT

To brief on issues and challenges, augmented reality (AR) developer needs to consider the technical and application issues that affect the user experience. The primary technological problems relating directly to the real-time, 3D interactive, arise from the thought of creating AR-based applications. Performance, interaction, and alignment are the three main technological issues in AR, and application issues consist of authoring and content creation. The other user experience affecting problems are as follows:

- User interface
- Compatibility with devices
- Visual interface
- Battery issues

An AR application is designed to operate in real time and may result in odd augmentation and not being compatible with the current state of the environment. Performance issues are unstoppable for an AR application when it comes to a device with limited power and processing speed. There should be a natural interaction between the user and the application in the real world without any breakpoints. The usability and the user experience will shatter if natural communication is disturbed. The same action and expertise should be observed when real-world objects interact with virtual objects. The camera pose and calibration need to be intact to maintain the augmented information at one position. If things don't go according to the plan with the device's camera, the virtual overlay will flutter out in the real world. This fluttering out effect tends to annoy people and is called an alignment error. The alignment error also depends on the user and camera position in the real-world environment. More explanation on camera position and augmentation errors is given in Chapter 4 of this book. The content creation is a primary aspect of AR application development and needs a database to extract visualizing information. For creating content in the proposed application, an online database manager called "Vuforia target manager" was used to upload markers and connected to unity for application development. The marker images were downloaded and imported as assets in the unity environment. Sometimes, for specific engines, the media format is unstable or invalid and needs to be converted into the system compatible format. Once the content is ready, there is an absolute necessity of getting it in the right shape and size to make it accurate with the user's expectations.

Besides content creation, authoring is a subject element of interest for developers and end-users. To speak in regard to the proposed application being an open-source, creation of the application should involve fewer programming aspects as the advanced AR technology is highly dependent on IDE connectivity. A nonexpert and minimal programming concept can allow people to combine objects and system communication at a conceptual level. As we discussed in the above sections, visual perceptions should support the application's purpose with a suitable type of rendering, i.e., photorealistic and nonphotorealistic rendering. The visual perception should help sustain the task without distracting the user experience and keep it smooth at maximum. The AR application should fall under certain guidelines and eligibility criteria after finishing the development process (Figure 6.1).

The diversification of mobile platforms is the main obstacle for wider use of the mobile application. The last point for this discussion is power consumption or battery life for the application usage. The application having its usage in the fields of retail stores and public places like restaurants and fitness centers must meet the "low power consumption" requirement because of the unavailability of charging stations at the application areas. The application requires the user to move freely, and thus, battery life is important. The AR application, which drains out the mobile device's battery in a few mins, is supposed to be unrealistic. There are other minor issues and challenges that are not discussed in detail in this chapter.

Other Challenges after Application Deployment

- **Network connection:** The AR application would want a constant network connection for marker recognition as the marker is stored on an online database, and fetching the details

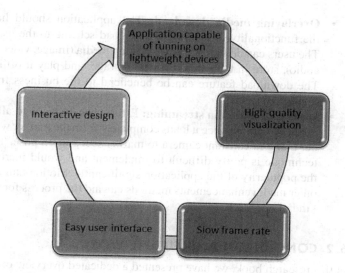

Figure 6.1 Challenges in AR application deployment.

from the internet would be required. There is an option of limiting the application to function without a network connection, but the data will be embedded in the application on download, and the package size would be ten times more than the one with internet requirement. Fortunately, in this digital era, there is internet availability on all public locations to access the AR application at no additional cost.

- **Instruction manual for using the application:** There are people of certain age groups who have adequate knowledge of using the internet and AR applications. The application should have a "tutorial" or "demo" embedded in the framework on how to use the application with buttons and widget mapping to let people know how to use the application.

- **Marker recognition on a busy location:** Another concern for using the AR application on public locations would be fast marker recognition. The marker should be recognized from a reasonable distance at one view and therefore, should be a strong marker. There should not be a situation where ten people gather at one location and wait for their chance to view overlaying information on their devices.

- **Overlaying media download:** The application should have the functionality of availing a download scheme to the users. The users can download the overlaying media (image, video, or audio), store in their devices local storage, and play it offline. The download feature can be beneficial to the business from financial aspects.
- **Undisturbed media streaming:** Efforts can be made to allow media streaming once it loads completely from the marker without being in constant camera to marker recognition pose. This technique is pretty difficult to implement and would increase the popularity of the application significantly. Motion blur and other media enhancements methods can aid the process for this kind of approach.

6.2 CONCLUSION AND FUTURE WORK

In this research book, we have presented a dedicated overview of AR concepts and application development techniques with the help of a solution manual. Marker-based tracking approach for the application has been the proposed topic of interest around the chapters in this book. Other core concepts for AR application design, system architecture, and workflow of the application with unity and Vuforia are described in the previous chapters of this book. We discussed the real-time use of AR system to help prevent the spread by following social distancing guidelines with the marker-based technique amidst COVID-19 pandemic to deploy a public health measure with the help of digital technology. The area of application considered for this typical use case is retail stores which avail free Wi-Fi or other network connectivity to their customers. Furthermore, the steps to develop the proposed application and screenshots and code are reported in this book.

We have presented how the author has contributed to several challenges in AR application development. In addition to problems, ways to improve and enhance the AR application, mitigation remedies, and public health measure solution for COVID-19 are reported in the later sections. The research can be concluded with the possibility of finetuning the proposed application and meeting the requirements for improved user experience. The application design can be expanded and extended for deployment in other areas such as fields of maintenance assembly, military, telecommunication, buildings and construction, interior design, public restaurants, and fitness centers. The research is scalable to any extent, where there is a huge amount of time and effort saved for humanity's betterment.

APPENDIX A
Sources for Getting AR Tools (FOSS)

1. **Unity**
    i. Unity Hub: https://public-cdn.cloud.unity3d.com/hub/prod/ UnityHubSetup.exe?_ga=2.142851573.155911673.16060 29116-1845815412.1591846015
    ii. Unity Editor: https://unity3d.com/get-unity/download/archive
2. **Vuforia**
    i. Add Vuforia Engine to a Unity Project or upgrade to the latest version: add-vuforia-package-9-5-4.unitypackage (2.57 KB)
    ii. Download for Android: vuforia-sdk-android-9-5-3.zip (22.04 MB)
    iii. Download for iOS: vuforia-sdk-ios-9-5-3.zip (61.71 MB)
    iv. Download for UWP: vuforia-sdk-uwp-9-5-3.zip (21.33 MB)
3. **Google ARCore**
    i. ARCore SDK for Android: arcore-android-sdk-1.21.0.zip
    ii. ARCore SDK for iOS: arcore-ios-sdk-1.21.0.zip
    iii. ARCore SDK for Unity: arcore-unity-sdk-1.21.0.unitypackage
    iv. ARCore Extensions for AR Foundation: arcore-unity-extensions-1.21.0.tgz
    v. ARCore SDK for Unreal: arcore-unreal-sdk-1.7.0.zip

APPENDIX B

Coding Guidelines for Developing and Deploying an AR App over an Android Phone

This step-by-step guide will teach you how to get started on building AR applications for Android using Unity and Vuforia.

DOWNLOAD AND INSTALL UNITY

Download here: https://store.unity.com/

Unity is a free piece of software that's typically used to create games. There are tons and tons of menus built into Unity which means that coding is kept to a minimum for your basic needs. Obviously as you go forward and build an actual app, you'll need to implement a fair amount of code.

BUILD AN AR APP USING VUFORIA

1. Download Vuforia SDK for unity and import it in Unity assets

Setting up Unity for AR Development

The next step is to set up the unity environment for augmented reality development.

1. In the GameObject dropdown menu, select "Vuforia > AR Camera." If a dialog box appears requesting for you to import additional assets, select "Import."

2. Next, in the GameObejct dropdown menu, select "Vuforia > Image" to add an Image Target to your scene.

To generate a development license key and create your free Vuforia account for your application, go to the Vuforia developer portal.

1. Select "Register" in the upper right corner.
2. Once your account is setup, select "Develop" in the menu panel and then click on "License Manager."
3. Click on "Get Development Key" on the License Manager page.
4. On the "Add a free Development License Key" page, name your application and agree to the terms and conditions.
5. To access your license key, return to the "License Manager" page in the Vuforia developer portal and click on the application name you just created. Copy the alphanumeric string, i.e., your License key on the right.
6. Return to Unity and select File > Build Setting and click on the "Player Settings" button at the bottom of the pop-up window.
7. Navigate back to the Inspector panel and click the "XR Settings" option located at the bottom of the menu. Now, select "Vuforia Augmented Reality Support" and accept the Vuforia license in Unity.
8. Next, select ARCamera in the Hierarchy pane and in the inspector panel, click on "Open Vuforia Configuration" button in the Vuforia Behavious (Script)" component. Copy and paste your license key in the App License Key text field.
9. Select "File > Save Scene as…" and give your scene a name. By default, Unity will save your scenes to the "Assets" folder of your project.

Convert Your Image Target to a Dataset

Now you will need a.jpg or.png image to create a dataset of your images on Vuforia. Using a visually complex image makes it easier for the device's camera to track the Image Target for your AR application.

You will need to upload your image to Vuforia to code it with tracking information.

1. Go to the Vuforia developer portal and select "Develop" in the top menu and then select "Target Manager."

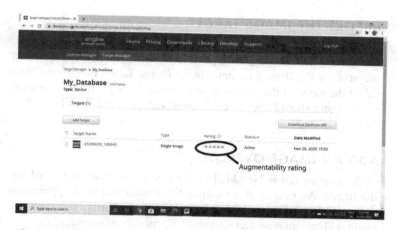

Figure B1.1 Image target Database created on Vuforia engine developer portal.

2. Click on the "Add Database" button on the Target Manager page. In the dialog box that appears, name your database and then select the device option. Click "Create."

3. Once your database has been created, return to the Target Manager page, and select your database. Select "Add Target" and choose "Single Image" as the type, add the height and width of your image, and upload by clicking on the "Browse" button (Figure B1.1).

Vuforia assigns target images a feature tracking rating on a scale of 1–5 stars. Navigate to the Image Targets database in the Vuforia Developer Portal and select the image you just uploaded. To reveal the image's trackable features, click on the Show Features link.

If your image is given a good rating (anything between 3 and 5 stars should work), access the Target Manager page, then select your image and click "Download Database." If you were creating an AR application with multiple Image Targets, you would want to convert all the images through Vuforia's developer portal before downloading them as a Unity package.

Select the "Download Database" button at the top right.

Select Unity Editor, and download the package. An Import dialog box will appear in Unity; click on "Import" to finish.

IMPORT THE IMAGE TARGET

In the project Hierarchy panel, select the ImageTarget GameObject. Next, navigate to the inspector panel, and then click on the drop-down menu of the Database parameter in the Image Target Behaviour. Now select the name of the database you created earlier. The ImageTarget game object should now be associated with your Database.

ADD AN IMAGE OVERLAY

An overlay has to be added to test if your AR camera is tracking the image. An overlay is a multimedia content that your application will superimpose onto your trigger image. In Unity, you can add audio, videos, images, and animated 3D models as overlays.

First, find an image that you want to use as an overlay. Drag the desired image into the "Assets" folder in the project of the Unity editor. Once the image has been successfully imported, select it, and change its Texture Type to "Sprite (2D and UI)" in the inspector panel. Click "Apply."

Next, drag the sprite you just imported onto your "Image Target" game object in the project Hierarchy.

The sprite is now a "child" of the Image Target game object and will only appear if the ARCamera recognizes the Image Target.

The position of your image overlay in the scene view is the same position at which it will appear when scanned with a mobile device camera. For example, if you place the overlay image at the top of the ImageTarget, then it will also appear at the top of the ImageTarget when your user scans it.

Next, you will need to adjust the size and position of your overlay image to center it on your ImageTarget using the translate, scale, and rotate tools located in the upper left of the Unity editor interface.

DEPLOYING THE APPLICATION

To install your own applications on your Android device,

1. Enable USB debugging by going to Setting > About Device.
2. Tap the Build number seven times.
3. Return to the previous screen, and you should now see a Developer Options tab. Click it and make sure the option for USB debugging is checked (Figure B1.2).

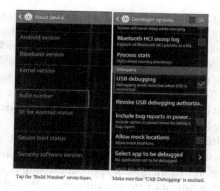

Figure B1.2 Steps to enable debugging mode on your Android device.

After all of your settings are set and your ImageTraget and overlay image are well transformed, you can proceed to your first build and run of the project. To be able to test your application, the easiest way is to have an Android device connected to your computer with Developer Mode and USB debugging enabled. This will allow you to deploy directly to your device and enable faster testing and feedback (Figure B1.3).

Open File > Build Settings.

You will first need to add a scene to the Build. Unity scenes can be added or removed at will from a build, and one of the easiest ways, in this case, is to just click Add Open Scenes. If this did not add the

Figure B1.3 Build settings for deploying AR application on Android device.

Main scene to the Scenes in Build list, then you can manually add it by finding it in your Assets and dragging and dropping it into the list.

We also want to make sure we have Android selected as the target platform. We can leave all of the other settings default for now. If you want to deploy to your device directly, make sure it is plugged in and select Build and Run. If everything was set up properly, Unity will take a bit of time to compile and build the application, and then attempt to deploy to your device.

With Android, it is very easy to share and test your completed application with other Android users without uploading it to the Google Play store. To share your application, simply send the.apk file as an email attachment to anyone with an Android device. However, before other users can download and install the.apk file, they will need to allow their Android device to install.apk files from non-Google Play sources by navigating to Settings > Security on their Android device and checking the box labelled "Unknown sources."

If everything was successful, the application will launch automatically, and you can test it out!

GLOSSARY

Aarogya Setu: Indian open source COVID-19 contact tracing app developed by the government.

Aesthetic: Philosophical terminology to describe nature and beauty with an emotional sense.

A.R: AR is a technology which stands for augmented reality and represents a virtual object in a 3D physical space.

ARCore: Google's ARCore is a software development environment, which allowes public to develop AR apps.

ARToolKit: Open source AR library deployed on Github to create AR applications.

Augmented Reality: Extended term for AR.

Canada COVID-19 Alert: COVID-19 alert is a digital contact tracing app developed by canadian government to tackle COVID-19.

Compatibility: Compatibility can be explained as a state where two objects or two physical elements fit together in one so as to generate an unbreakable flow.

Containment: Action in which someone's or some object's power in any form is kept limited to or within a certain boundary.

COVID-19: Infectious disease which is commonly spread contagiously, and the common symptoms include fever, headache, vommiting, etc.

Design: Design can be defined as a layout or overview which displays the working of a system before it has been built.

Diversification: The process of enlarging a particular domain by embedding a wide range of products in the same field of operation.

DROR: Another digital contact tracing app developed by Indian citizen to fight COVID-19.

Endemic: Endemic is a situation where a disease spread is found in a different community of people in a certain area.

Epidemic: Spread of a dangerous disease in a specific community at a particular time.

GPS: GPS stands for Global Positioning System, which is a satellite-based radionavigation system owned by the United States government.

HCI: Human Computer Interaction (HCI) is a field of study which focuses on the technology interface interaction between humans and computer.

HMDs: Head-mounted displays.

IDE: IDE stands for Integrated Development Environment, which is also a platform to conduct coding and development practices.

Invasion: Invasion can be simply defined as a situation of an unwelcomed intrusion in someone else's domain.

Image Construction: A technological term in which the image is constructed by an algotihm running in the system and then preprocessed.

Image Detection: Image detection is a process in which the image is detected with the help of a camera object attached to the system.

Image Processing: A technological term in which the image is processed after being constructed and is passed through cleaning, transformation, reduction, and postprocessing later.

Image Recognition: A technological term in computer vision field where the image is recognized of a class by a camera object.

Image Rendering: Image rendering is a process of generating a realistic or nonrealistic image from a 2D/3D model with the help of a computer.

Imposition: The act or a process of burdening someone with a set of rules as a demand or a forceful act.

Interactive: The act of communicating with someone or blending in an environment with someone to exchange a set of words.

Localization: The concept of becoming located or fixed at a particular place at a particular time.

M.A.R: M.A.R stands for mobile augmented reality in which augmented reality-based applications are of shere focus.

Marker-based AR: Derived from AR, marker-based AR is a technique in which augmented information is triggered upon scanning a marker.

Marker-less AR: Dervied from AR, markerless AR is a technique is which augmentation is triggered upon location basis. GPS in particular.

Mitigation: Mitigation can be defined as an act of reducing the severity or seriousness of something.

Mixed Reality: Mixed reality is a merger of augmented and virtual relaity technologies which has its own set of applications.

Model Construction: The phenomenon in which the model of a system is constrcuted on the basis of methodology elements.

Monitoring: Monitoring is a process of observing something for a longer period of time with the intention of investigation.

Odometry: Use of data from motion sensors to estimate change in position over time is termed as odometry.

Outliers: Outliers in a system can be defined as substances or elements which are away or detached from the main system.

QR-code: A machine-readable label that contains some data like a link or digital trigger.

Remedies: Remedies can be defined as treatment options or measures to tackle a problem.

Rendering: Rendering is a type of image synthesis process in computer graphics which generates a 2D or 3D model by means of computer.

SARS: SARS (Severe Acute Respiratory Syndrome) is a viral respiratory illness caused by dangerous virus.

SDK: SDK stands for software development environment toolkits such as android SDK, virtual reality SDK, etc.

Self-Isolation: Self-isolation pr isolating is a process in which an individual is kept in a restricted environment without any touch from external environment.

Simulation: Simulation in science that can be defined as imitating a process on a local machine with the help of assistive tools and technology

Social Distancing: Social distancing is a practice in which the human beings are supposed to mainatin a 6 feet distance from each other in a any public space.

SODAR: SODAR is an AR application developed and launched by google to visualise social distancing guidelines around you.

TAM: Technology Acceptance Model (TAM) is an IT system theory that models how users come to accept and use technology.

Unity Engine: Unity is a cross-platform engine developed by Unity technologies and real-time games, AR apps can be created on this platform.

User Interface: User interafce can be defined as a front-end human interaction platform where end-users can interact with the database objects and retreive infromation by querying or clicking.

Visualization: Visualization can be termed as a representation of data or a set of information in form of charts and other diagrammatic representations.

Vuforia: Vuforia is a SDK/plugin environment which makes it easier to create cutting-edge augmented reality applications.

WHO: The World Health Organization is a specialized agency of the united nations responsible for international public health.

REFERENCES

1. Last, J. M., (ed.) (2001). *Dictionary of epidemiology.* 4th ed. New York: Oxford University Press, p. 61.
2. Center for Disease Control. (2011). *Principles of epidemiology in public health practice*, Chapter 1, Book originally published: October 2006, Book updated: November 2011, U.S. Department of Health and Human Services, Centers for Disease Control and Prevention (CDC) Office of Workforce and Career Development Atlanta, GA.
3. Anderson, R. M., Heesterbeek, H., Klinkenberg, D., & Hollingsworth, T. D. (March 2020). How will country-based mitigation measures influence the course of the COVID-19 epidemic? *Lancet*, 395(10228), 931–934. doi:10.1016/S0140-6736(20)30567-5. PMC 7158572. PMID 32164834.
4. Azuma, R. T. (August 1997). A survey of augmented reality. In *Presence: Teleoperators and Virtual Environments*, 6(4), 355–385.
5. Crain, D. W. "TV Object locator and image identifier," US Patent 4,084,184.
6. Berlin, L. (11 July 2009). "Kicking reality up a Notch," *The New York Times.* ISSN 0362-4331
7. Sanders, M. S., & McCormick, E. J. (1993). *Human factors in engineering and design* (7th ed.). New York: Mcgraw-Hill.
8. Baek, A., Lee, K., & Choi, H. (2013). "CPU and GPU parallel processing for mobile Augmented Reality," *2013 6th International Congress on Image and Signal Processing (CISP)*, Hangzhou, 2013, pp. 133–137, doi: 10.1109/CISP.2013.6743972.
9. Bonfiglio, N. (1 October, 2018). "DeepMind partners with gaming company for AI research," *The Daily Dot.*
10. Matney, L. (25 May, 2017). "With new realities to build, Unity positioned to become tech giant," *TechCrunch.*
11. Jung, T., & Claudia Tom Dieck, M. (4 September, 2017). *Augmented reality and virtual reality: Empowering human, place and business.* Cham, Switzerland: Springer International Publishing.

12. Huang, J., Kinateder, M., Dunn, M. J., Jarosz, Y., & Cooper, E. A. (16 January, 2019). An augmented reality sign-reading assistant for users with reduced vision. *PLoS One*, 14(1), e0210630. doi:10.1371/journal.pone.0210630. eCollection 2019.

13. Lv, Z., Halawani, A., Feng, S., et al. (2015). Touch-less interactive augmented reality game on vision-based wearable device. *Pervasive Ubiquitous Computing*, 19, 551–567. doi:10.1007/s00779-015-0844-1.

14. DePaolis, L. (30 July, 2018). Augmented reality and Myo for a touchless interaction with virtual organs. http://avrlab.it/augmented-reality-and-myo-for-a-touchless-interaction-with-virtual-organs/

15. Zvejnieks, G. (18 October, 2018). Marker-based vs markerless augmented reality: Pros, cons & examples. https://overlyapp.com/blog/marker-based-vs-markerless-augmented-reality-pros-cons-examples/

16. Morris, S., Fawcett, G., Brisebois, L., & Hughes, J. (28 November, 2018). A demographic, employment and income profile of Canadians with disabilities aged 15 years and over, 2017. https://www150.statcan.gc.ca/n1/pub/89-654-x/89-654-x2018002-eng.htm

17. Jung, T., Claudia Tom Dieck, M., & tom Dieck, D. (2014). A theoretical model of augmented reality acceptance. *e-Review of Tourism Research*, Vol. 5.

18. Ayeh, J., Au, N., & Law, R. (2013). "Towards an understanding of online travellers. Acceptance of consumer-generated media for travel planning: Integrated technology acceptance and source credibility factors." In: Cantoni L., Xiang Z. (eds) Information and Communication Technologies in Tourism 2013. Berlin, Heidelberg: Springer. https://doi.org/10.1007/978-3-642-36309-2_22

19. Venkatesh, V., & Bala, H. (2008). Technology acceptance Model 3 and a research agenda on interventions. *Decision Sciences*, 39(2), 273–314.

20. Amoako-Gyampah, K., & Salam, A. (2004). An extension of the technology acceptance model in an ERP implementation environment. *Information and Management*, 41, 731–745.

21. Herbst, I., Braun, A., McCall, R., & Broll, W. (2008). TimeWarp: Interactive time travel with a mobile mixed reality game. Paper presented at *MobileHCI 2008*, Amsterdam, The Netherlands.

22. Kounavis, C. D., Kasimati, A. E., & Zamani, E. D. (2012). Enhancing the tourism experience through mobile augmented reality: Challenges and prospects. *International Journal of Engineering Business Management*, 4(10), 1–6.

23. Lin, C., Shih, H., & Sher, P. (2007). Integrating technology readiness into technology acceptance: The TRAM model. *Psychology and Marketing*, 24(7), 641–657.

24. Hirzer, M. (2008). *Marker detection for augmented reality applications*. Austria: Inst. for Computer Graphics and Vision, Graz University of Technology.

25. Li, S., & Xu, C. *Efficient lookup table based camera pose estimation for augmented reality*. Wiley Online Library, https://doi.org/10.1002/cav.385.

26. Cheng, Jack C. P., Chen, K., & Chen, W. (2017). "Comparison of marker-based A.R. and markerless AR: A case study on indoor decoration system," In *Proc. Lean & Computing in Construction Congress* (L.C. 3), Vol. 2 (CONVR), Heraklion, Greece.

27. Oufqir, Z., El Abderrahmani, A., & Satori, K. (2020). "From Marker to markerless in augmented reality," In Bhateja, V., Satapathy, S., & Satori, H. (eds.), *Embedded systems and artificial intelligence. Advances in intelligent systems and computing*, vol. 1076. Singapore: Springer.

28. Unity3D, what is Unity? Available at http://goo.gl/n3YtYX, (Accessed 20 June, 2020).

29. Vuforia User Manual. Available at https://docs.unity3d.com/2019.1/Documentation/Manual/vuforia_configuration.html, (Accessed 21 June, 2020).

30. Web-source reference- https://scandinavian.ca/

31. Choudhary, M. How IoT can help fight COVID-19 battle. https://www.geospatialworld.net/blogs/how-iot-can-help-fight-covid-19-battle, (Accessed 3 June, 2020).

32. 18 Most Popular IoT devices in 2020 (Only Noteworthy IoT Products). https://www.softwaretestinghelp.com/iot-devices/, (Accessed 3 June, 2020)

INDEX

Printed in the United States
by Baker & Taylor Publisher Services

Printed in the United States
by Baker & Taylor Publisher Services